THE THEORY OF EVERYTHING

The Origin and Fate of the Universe

宇宙简史

起源与归宿

[英国]斯蒂芬·霍金 Stephen W. Hawking 著

赵君亮 译

译林出版社

图书在版编目（CIP）数据

宇宙简史：起源与归宿／（英）霍金（Hawking, S. W.）著；赵君亮译．—南京：译林出版社，2012.5（2023.3重印）
书名原文：The Theory of Everything: The Origin and Fate of the Universe
ISBN 978-7-5447-2786-0

Ⅰ.①宇…　Ⅱ.①霍…②赵…　Ⅲ.①宇宙学－普及读物　Ⅳ.①P159-49

中国版本图书馆 CIP 数据核字（2005）第 082324 号

The Theory of Everything: The Origin and Fate of the Universe
by Stephen William Hawking

著作权合同登记号　图字：10-2018-426号

宇宙简史：起源与归宿［英国］斯蒂芬·霍金／著　赵君亮／译

责任编辑　陈　叶
责任印制　单　莉

原文出版　Phoenix Books，2007
出版发行　译林出版社
地　　址　南京市湖南路 1 号 A 楼
邮　　箱　yilin@yilin.com
网　　址　www.yilin.com
市场热线　025-86633278
排　　版　南京展望文化发展有限公司
印　　刷　江苏凤凰新华印务集团有限公司
开　　本　718 毫米 × 1000 毫米　1/16
印　　张　13.75
插　　页　2
版　　次　2012 年 5 月第 1 版
印　　次　2023 年 3 月第 29 次印刷
书　　号　ISBN 978-7-5447-2786-0
定　　价　39.00 元

译者前言

斯蒂芬·W.霍金1942年1月8日出生于英国牛津，这一天恰好是伽利略逝世三百周年忌日。他曾先后毕业于牛津大学和剑桥大学，1965年获剑桥大学哲学博士学位。在学生时期的1963年，二十一岁的霍金被诊断患有肌肉萎缩症，即卢伽雷氏症，自此他便终日与轮椅为伴。1985年后更因丧失语言能力，与人交流只能通过语音合成器来完成，为合成一小时录音演讲往往需要准备十天时间。尽管身残如此，霍金依然头脑清晰而又聪慧，在量子力学、宇宙学，特别是有关黑洞的研究中取得了一系列辉煌的成就，被人们誉为世界上最杰出的天才之一，现今在世的最伟大科学家，当代的爱因斯坦。他获得过多项荣誉及奖励，如1974年当选为英国皇家学会最年轻会员，1974—1975年成为美国加州理工学院费尔柴尔德讲座功勋学者，1978年荣获世界理论物理研究最高奖——阿尔伯特·爱因斯坦奖，1989年获得英国的爵士荣誉称号。

霍金在他长达四十余年的科学生涯中建树颇多。他与彭罗斯一起，在广义相对论框架内证明了著名的奇点定理，为此两人共同获得1988年沃尔夫物理奖。霍金对黑洞的关注长达十余年之久，他证明了黑洞的面积定理，并发现黑洞会像黑体一样发出辐射，而辐射的温度与黑洞质量成反比，这就是今日人所共知的“霍金辐射”。

霍金不仅是一位罕见的杰出科学家，而且是一位广受人们欢迎的科普学者。他深信科学理论进展的一般性原理应该能为广大公众所理解，他的演讲在国际上享有盛誉，他的足迹遍布世界各地，其中包括曾多次来中国讲演。在霍金的诸多科普著作中,最有名的当推《时间简史》。该书已被译成四十余种文字,出版数超过一千万册。霍金试图通过自己的书籍和通俗演讲,把自己的科学思想与全世界各阶层人士进行广泛的交流。

本书是霍金的又一力著,全书由七次相对独立但又互相紧密联系的讲座集结而成。作者力图通过这一系列讲座,用尽可能通俗的语言,深入浅出地阐明人类迄今所探知的宇宙演化史的基本轮廓,以及相关理论的基本原理。读者通过阅读本书,可以初步认知宇宙从形成、演变直至可能的最终归宿之全过程,以及这一过程之所以发生的物理机制。

书中涉及了现代物理学的诸多领域,如宇宙的大爆炸起源,黑洞的特性,以及时空本质等,为此作者就讲座次序及其相关内容做了精心安排。在宇宙学方面,从介绍两千多年前亚里士多德时代人们对地球的朦胧认识,到今天颇为深奥的大爆炸宇宙论和暴胀理论。在引力理论方面,从早期的牛顿引力论到爱因斯坦广义相对论,进而融入现代量子理论,甚至还简要叙述了有关弦理论的种种趣事。黑洞性质的阐述无疑是本书的重点之一,为此作者特意安排了整整两讲的篇幅,并穿插若干富于想象力的精彩描述,使人阅后难以忘怀。在讨论了宇宙的起源与归宿之后,最后一讲的内容回归到本书的主旨——“万物之理”,即如何寻求一种统一理论,以能囊括所有的局部性物理学相互作用,而这正是现代物理学尚未解决的若干极其重要问题之一。作者认为一旦做到这一点,我们将会真正理解宇宙以及人类在宇宙中所处的地位。

可以说,本书就人类认识宇宙的历史和现状,向读者展现了一次逐步深入的探索式“旅行”。实际上,在这一认识过程中,必然涉及相当深奥的物理学和天文学理论,然而在全书中仅有一处出现简单公式(即$E=MC^2$),由此足见作者不仅是一代科学大师,而且还是向公众传播科学知识和科学方法的巧匠。尽管个别章节对一些读者来说尚有一定难度,但这并不会影响读者对作者总体思想的认知和把握。如有读者感到此类难点,尽可跳过无妨,重要的是了解宇宙如何诞生和演化的历程,以及所涉及的一些重要理论之基本概念。

囿于译者各方面的学识水平,译文的错误和疏漏之处在所难免,还望读者不吝赐教为盼。

赵君亮

2009年5月13日

目录

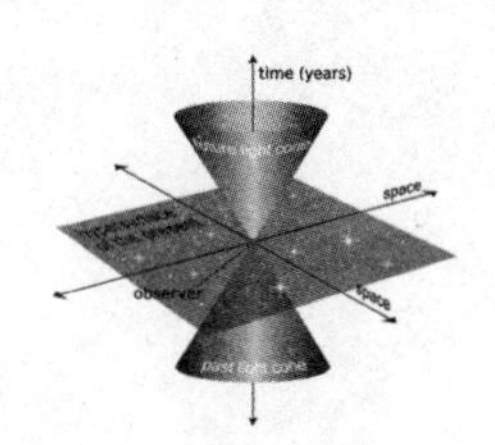

引言

我想尝试通过这一系列讲座，来阐明我们所认识的宇宙史之基本轮廓，从大爆炸到黑洞。在第一讲中我将简要回顾过去年代中有关宇宙的若干观念，以及如何取得目前的宇宙图像。您也许可以把这称为宇宙史的历史。

在第二讲中我要说明的问题是，如何从牛顿和爱因斯坦的两种引力理论推断，宇宙不可能是静态的；它只能处于膨胀或收缩之中，两者必居其一。接下来的推论是，在100亿到200亿年之前的某个时间，宇宙的密度为无穷大。这一时刻称为大爆炸，宇宙应该由此诞生。

第三讲的内容与黑洞有关。当一颗大质量恒星，或者质量更大的天体，在自引力作用下发生坍缩[1]，便可以形成黑洞。根据爱因斯坦的广义相对论[2]，如果有人愚不可及，一步跌入黑洞，那就会一去不返，永远消失。他们绝不可能再度从黑洞中逃逸出来。相反，对这样的人来说经历会是非常痛苦的，等待着他们的是最终到达一个奇点。但是，广义相对论是一种经典理论——这就是说，它没有考虑到量子力学的测不准原理[3]。

在第四讲中我将要阐明量子力学如何允许能量从

① 坍缩：在满足一定的前提条件下，天体会在自引力作用下向内收缩，直至达到新的平衡态，这一过程称为坍缩，如星云经坍缩而形成恒星，恒星坍缩成白矮星等。（本书所有边注均为译者所加）

② 广义相对论：研究物质在空间和时间中如何进行引力相互作用的理论，与牛顿引力理论完全不同，由美国物理学家爱因斯坦于1915年完成。

③ 测不准原理：量子力学的一条重要原理，说明微观客体的坐标和动量不可能同时具有确定的值，由德国物理学家海森伯于1927年提出，又称“测不准关系”。

阿尔伯特·爱因斯坦
(1879—1955)

黑洞中逸出，黑洞并非人们所描绘的那样黑不可知。

第五讲的主题涉及把量子力学的一些观念用于大爆炸和宇宙起源。由此引出的重要概念是，时空可以在范围上是有限的，但没有边际，或者说它是无界的。这有点像地球的表面，不过还得增加两维。

第六讲中所要讨论的问题是，尽管物理学定律就时间上来说是对称的，但这种新的、有关宇宙边界的设想，也许可以用来解释过去和未来为什么会有霄壤之别。

最后，第七讲中我要阐述的是，人们如何尽力寻求一种统一理论，以能囊括量子力学、引力以及物理学中的其他所有相互作用。一旦做到这一点，我们将会真正理解宇宙以及我们在宇宙中的地位。

大质量恒星坍缩后形成的蟹状星云

第*1*讲

有关宇宙的若干观念

早在公元前340年,亚里士多德在他的《天论》(*On the Heavens*)一书中,就已能提出两个令人信服的论据,从而证明地球是一个圆球,而不是一个扁平的盘。首先,他领悟到月食的成因是由于地球运行到了太阳和月球之间。地球投射在月球上的影子始终是圆形的,这一观测事实只有当地球为球形时才能出现。如果地球是一个扁平的圆盘,那么除非月食发生之际太阳总是正射到盘的中心,否则地球的影子必然会拉长而成为椭圆形。

亚里士多德(前384—前322)

第二,希腊人在他们的迁居过程中早就知晓,南方看到的北极星在天空中的位置,要比北部地区所看到的位置来得低。根据埃及和希腊两地所见北极星视位置的差异,亚里士多德甚至估计出了地球的周长为40万斯达地[①]。1斯达地的准确长度无人知晓,不过很可能约合200码。如是,亚里士多德的估值约为目前所采用值的两倍。

① 斯达地(Stadium)系古希腊长度单位,约等于185米。

此外,希腊人还提出了地球必然为球形的第三个理由:为什么船舶出现在地平线上时,人们先看到的是船帆,然后才看到船身?亚里士多德认为地球是静止不动的,而太阳、月球、行星以及恒星都沿着圆形轨道绕地球运动。他深信,鉴于某些神秘莫测的理由,地球位于宇宙的中心，而圆运动是最完美的。

公元2世纪[①]，经托勒密的精心推敲，形成了一种完整的宇宙模型。地球位于中心，它的周围有八个天球，分别承载了月球、太阳、恒星以及当时所知道的五颗行星，即水星、金星、火星、木星和土星。这些行星分别在一些较小的圆轨道上运动，而这些小圆圈又各自附于上面提到的那些天球上，由此来说明所观测到的行星在天空中的复杂运动路径。位于最外面那个天球上的则是一些所谓固定不动的恒星，它们之间的相对位置始终保持不变，同时又作为一个整体在天空中转动。至于在最外面的天球之外又是什么，则从来没有被搞清楚过，但这肯定不是人类可观测的宇宙的组成部分。

① 原文为“公元1世纪”，有误。托勒密于公元140年在《天文学大成》一书中完整地提出了他的地心说。认为地球位于宇宙的中心且静止不动，而其他所有天体都绕着地球转动。

托勒密 (100—170)

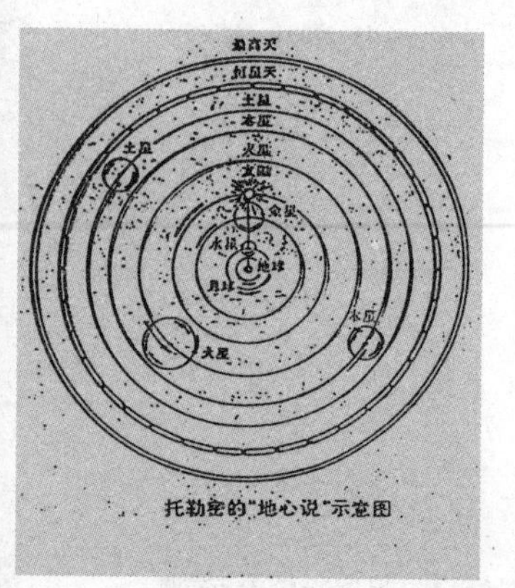

托勒密的“地心说”

托勒密的模型提供了一种颇为合理的精确系统，它可以用来预测诸天体在天空中的位置。但是，为了正确地预测这些位置，托勒密不得不设定月球运动的路径距地球时近时远，最近时的月地距离只是其他时候的一半。这意味着有时候月球看上去会比通常所看到的大上一倍。托勒密本人知道这是一个问题，但尽管有这个缺陷，他的模型在当时为大多数人所接受，虽然并非人人都予以认可。基督教教会接纳了托勒密的模型以作为宇宙的图像，因为它与《圣经》的记载相符。该模型的一大好处是，它在恒星球之外为天堂和地狱留下了广阔的空间。

然而，波兰教士尼古拉·哥白尼在1514年提出了一种更为简单的模型。最初，哥白尼因担心会被指控为异端之说，便采用匿名方式发表了他的模型。哥白尼的思想是，太阳位于中心且静止不动，地球和其他行星都绕着太阳在圆轨道上运动。对哥白尼来说可悲的是，差不多在一个世纪之后人们才认真地接受了他的思想。其时，有两位天文学家，德国人约翰内斯·开普勒[①]和意大利人加利莱奥·伽利略——开始公开支持哥白尼的理论，尽管事实上该理论所预言的行星运动轨道与观测结果并不完全相符。亚里士多德—托勒密理论的消亡始于1609年。在那一年，伽利略开始用望远镜观测夜空，当时望远镜才刚发明不久[②]。

① 开普勒行星运动定律：17世纪初由德国天文学家开普勒提出，共有三条：所有行星的运行轨道都是椭圆，太阳位于椭圆的一个焦点上；行星与太阳的连线在相等的时间内扫过的面积相等；行星公转运动周期的平方与行星公转轨道半长径的立方成正比。

开普勒 (1571—1630)

② 望远镜的发明：1608年荷兰眼镜制造商利伯希在一次偶然机会中发明了望远镜；翌年，伽利略在得知此消息后亲手制作了望远镜并用于观测天体，做出了一系列重大发现，从而开辟了天文观测的新纪元。

伽利略在观测木星时，发现它的周围有几颗小的卫星（或者说月亮）在绕着木星作轨道运动。这说明所有天体并非如亚里士多德和托勒密所认为的那样都必然直接绕着地球运动。当然，仍然可以相信地球位于宇宙中心且静止不动，不过这时要使木星的月亮看上去表现为在绕木星运动，那么它们绕地球运动的路径必然极其复杂。然而，哥白尼理论就要简单得多了。

在同一时期，开普勒对哥白尼理论作了修正，他认为行星运动的轨道不是圆，而是椭圆。这么一来，理论预期与实测结果最终完全相符了。就开普勒而言，椭圆轨道只是一种特定的假设，而且是一种颇不受人欢迎的假设，因为在那时的人看来椭圆显然不如圆来得完美。开普勒发现椭圆轨道与观测结果很好地相符带有某种偶然性，他当时认为由于磁力的作用才使得行星绕太阳运动，而椭圆轨道与这种观念是无法调和的。

伽利略的折射望远镜

只是到了多年后的1687年，牛顿才对此给出了解释，那一年牛顿发表了他的名著《自然哲学的数学原理》。这本书也许是物理科学领域迄今为止所出版的一部最为重要的著作，书中牛顿不仅提出了描述物体在空间和时间中运动规律的理论，而且还推导出了分析这类运动所需要的数学公式。不仅如此，牛顿还提出了万有引力定律。这条定律指出，宇宙中的每一个物体都会受到其他所有物体的吸引，物体的质量越大，物体间的距离越近，引力就越强。正是因为这种作用力的存在，才使得物体会落到地面上来。关于一个苹果掉到牛顿头上的故事似乎并不足以为信。牛顿本人提到过的仅仅是，关于引力的思想是他处于沉思冥想之际，由一个苹果的掉落而引发的。

存在第一推动力吗？

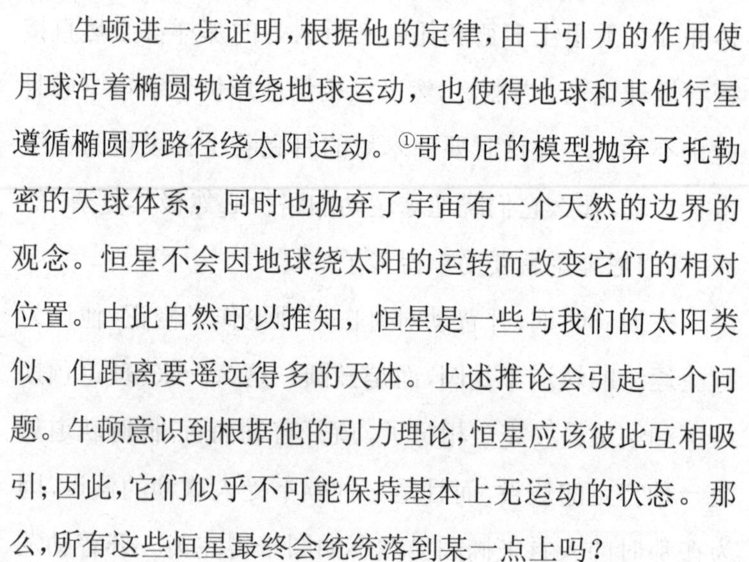

牛顿进一步证明，根据他的定律，由于引力的作用使月球沿着椭圆轨道绕地球运动，也使得地球和其他行星遵循椭圆形路径绕太阳运动。①哥白尼的模型抛弃了托勒密的天球体系，同时也抛弃了宇宙有一个天然的边界的观念。恒星不会因地球绕太阳的运转而改变它们的相对位置。由此自然可以推知，恒星是一些与我们的太阳类似、但距离要遥远得多的天体。上述推论会引起一个问题。牛顿意识到根据他的引力理论，恒星应该彼此互相吸引；因此，它们似乎不可能保持基本上无运动的状态。那么，所有这些恒星最终会统统落到某一点上吗？

① 晚年的牛顿在研究行星为什么会围绕太阳运转时，由于信奉上帝，认为除了有引力的作用外，还有一个“切线力”，该“切线力”只能是来自上帝的“第一推动力”。

牛顿在1691年写给当时另一位权威思想家理查德·本特利的一封信中指出，如果仅有有限数目的恒星，上述情况确实是会发生的。但是另一方面，他又推断说，如果恒星的个数为无穷大，且又大致均匀地分布在无限大的空间内，那么这种情况就不会出现，因为这时对恒星来说就不存在任何使之内落、集聚的中心点。这种推论是人们谈论关于无限的问题时可能遭遇的陷阱的一个例子。

在一个无限的宇宙中，每一个点都可以被视为中心，因为在每一点处朝各个方向看去都会有无穷多颗恒星。只是在多年之后人们才领悟到，认识这一问题的正确途径是，应该考虑的是一种有限的空间，其中的恒星都会彼此内落并集聚。现在我们要问，如果在上述有限区域的外围加上一些恒星，且它们大体上为均匀分布，那么情况会有哪些变化？根据牛顿定律，后来补充的恒星与原来的那些恒星毫无区别，因而它们也会接连不断地内落。这样的恒星可以想增加多少就增加多少而不受限制，但它们会始终保持不断地自行坍缩。现在我们知道了，不可能构筑一个静态的无限宇宙模型，在其中引力永远是一种吸引力。

在20世纪之前从未有人提出过宇宙正处于膨胀或收缩之中，这耐人寻味地反映了当时的主流思潮。当时为人们普遍接受的观念是，宇宙要么从来就以一种不变的状态永恒存在，要么它是在过去某个确定的时刻被创生出来，而且宇宙诞生时的状态与今天所观测到的样子大体上是一样的。形成这种观念的部分原因也许在于，人们倾向于相信永恒的真理，以及从下述想法中所得到的些许安慰：尽管他们会慢慢地老去，直至死亡，但宇宙是永恒不变的。

即使有人意识到牛顿引力理论表明宇宙不可能是静

态的，他们也不会去思考并提出宇宙也许正处于膨胀之中。相反，他们尝试去修正引力理论，办法是在很大的距离上使引力变为斥力。这种做法不会对预测行星的运动产生显著影响。但是，它可以使无限分布的恒星保持平衡状态：近距离恒星间的引力被来自远距离恒星的斥力相抵消。

然而，我们现在认为这种平衡态是不稳定的。如果某一天区内的恒星哪怕只是彼此间稍稍靠近一点儿，它们之间的吸引力就会增强，并超过斥力的作用。这意味着那些恒星便会继续彼此内落、集聚。另一方面，要是恒星之间的距离略有增大，斥力就会占上风，结果使恒星进一步互相远离。

人们通常认为，对无限静态宇宙的另一个反诘是由德国哲学家海因里希•奥伯斯提出来的。事实上，与牛顿同时代的各行各业学者已经提出了这个问题，甚至奥伯斯1823年的文章也不是包含了对这一议题貌似合理的推论的第一篇。不过，这是最早受到人们广泛关注的一篇文章。困难之处在于，在一个无限静态的宇宙中，几乎每一条视线或者每一条边，都将终止于某颗恒星的表面。因此，人们应当看到整个天空会像太阳一样明亮，哪怕在夜晚也是如此。奥伯斯对此的解释是，来自遥远恒星的光线因受到行进路径上物质的吸收而减弱了。但是，如果情况确实如此，那么这类介质也会因受到加热而发光，并最终变得如恒星般明亮。

为了避免得出整个夜间天空会变得如太阳表面那样

明亮的结论，唯一的途径是假定恒星并非永远在发光，它们只是从过去某个确定的时刻起才开始发出光芒。在这种情况下，起吸收作用的介质也许迄今尚未得以充分加热，或者遥远恒星所发出的光线可能尚未到达我们这里。这就会引出另一个问题：是什么原因能使恒星在原初位置上开始发光？

宇宙之开端

诚然，有关宇宙之开端的讨论可谓是由来已久。在犹太教、基督教或伊斯兰教的早期传说中有着若干种宇宙学，根据这类宇宙学，宇宙应始于过去某个有限而并不太遥远的时刻。之所以存在这样一个开端的一个理由是，感觉上必然要有一个造物主来解释宇宙的存在。

另一个论点由圣奥古斯丁在他的《上帝之城》(*The City of God*)一书中提出。圣奥古斯丁指出，文明的发展是渐进式的，而我们记住了是谁完成了这项业绩，又是谁开发出了那项技术。有鉴于此，人类——因而也许还有宇宙，就不可能已经存在了太长的时间。不然的话，今天人类文明的进展应当比我们现已取得的更为超前。

圣奥古斯丁 (354—430)

依据《创世纪》一书所述，圣奥古斯丁所采用的宇宙创生之时约为公元前5000年。有意思的是，这一时间距最近一次冰河期结束之际不算太远，该冰期约终结于公元前10000年，而那时人类文明已经萌发了。另一方面，亚里士多德和大多数其他希腊哲学家并不喜欢创生的观念，因为这掺入了太多的神授因素。所以，他们认为人类和人类周围的世界在过去和将来都是永恒存在的。他们已经考虑到了前面所提到的关于文明进展的论点，对此他们的辩答是，由于洪水和其他天灾的周期性出现，人类一次又一次地退回到文明的起端。

当大部分人对一个基本上处于静态、无变化的宇宙深信不疑之时，宇宙是否有一个开端的问题，实质上便成了某种玄学或神学问题。人们可以就两条不同的途径来说明所观测到的现象：或者宇宙永恒存在，或者它在某个有限时间内处于运动之中，而运动的方式恰好使宇宙看上去就像是永恒存在一样。但是在1929年，埃德温·哈勃[①]完成了一项划时代的观测，即无论你朝何处看，遥远的恒星[②]都在快速地远离我们而去。换言之，宇宙正在膨胀。这意味着在过去的某个时间，天体应该紧密地集聚在一起。

事实上，似乎在大约100亿或200亿年前的某个时间，所有这些天体都恰好位于相同的位置上。

这一发现最终把宇宙之开端的问题纳入了科学的范畴。哈勃的观测表明，曾经存在一个称之为大爆炸的时刻，那时宇宙为无限小，因而其密度必为无穷大。如果在这之前还曾出现过一些事件，那么这类事件也不会影响到现在所发生的一切。它们的存在可以忽略而不予考虑，因为它们不会产生任何观测效应。

人们可以说时间有一个起点，即大爆炸瞬刻，这意味着在这之前的时间是完全不可定义的。应该强调的是，时间有起点之说与以前习以为常的观念大不相同。在一个

① 以天文学家哈勃为名的哈勃望远镜是世界上最大、图像最清晰的天文望远镜。1990年4月5日由美国“发现号”航天飞机送入太空，总重量约11.5吨，望远镜口径2.4米，造价20亿美元，是迄今取得科学成果最多的空间天文项目。

② 原文如此。这里恰当的用词应当是“星系”，而不是“恒星”。

无变化的宇宙中，时间上的起点必然是由来自宇宙之外、某种不为人知的外因所赋予的。对于一个起点来说，并不存在物理学上的必然性。人们可以设想，上帝确实在过去的任意时刻创造出了宇宙。另一方面，如果宇宙正在膨胀，那么也许存在一些物理学上的理由，可用来说明为什么必然有过一个开端。人们仍然可以相信，是上帝在大爆炸瞬间创造出了宇宙。上帝甚至可以在大爆炸后的某个时刻创造出宇宙，不过创造的方式恰好能使宇宙看上去曾经历过一次大爆炸。但是，设定宇宙创生于大爆炸之前是毫无意义的。一个膨胀中的宇宙并不排斥创世主的存在，但它确实对创世主有可能完成其使命的时间划定了范围。

哈勃望远镜

第2讲
膨胀的宇宙

银河系是一个庞大的恒星系统，而我们的太阳以及邻近的恒星全都是银河系的组成部分。长期以来，人们一直以为银河系就是整个宇宙。只是到了1924年，美国天文学家埃德温•哈勃才证实我们的星系并不是独一无二的。事实上，还存在着许许多多其他的星系，而在星系之间则是广袤的虚无空间。为了证明这一点，哈勃必须确定这些河外星系的距离。我们可以确定邻近恒星的距离，办法是观测它们因地球绕太阳运动而引起的位置变化。但是，河外星系实在是太过遥远了，这与近距离恒星的情况不同，它们看上去完全固定不动。因此，哈勃只能通过间接的方法来测量它们的距离。

须知，恒星的视亮度取决于两个因素：光度，以及它离我们有多远。对于近距离恒星来说，我们可以测得它们的视亮度和距离，于是便能确定它们的光度。相反，要是我们知道了其他星系中一些恒星的光度，就可以通过测定它们的视亮度来推算出它们的距离。哈勃论证了存在某些类型的恒星，当它们距离我们近得足以被我们测量时，它们有相同的光度。于是，如在另一个星系中发现了同类恒星，我们就可以设想它们有着同样的光度。这样一来，便可以计算出那个星系的距离。如果可以对同一个星系中的若干颗恒星实施此类计算，并总是得出相同的距离，那么对星系距离的估计就相当可信了。通过这条途径，哈勃得到了九个不同星系的距离。

现在我们知道，利用现代望远镜可以观测到数千亿个星系，银河系只是其中之一，而每个星系又含有数千亿颗恒星。我们生活在一个缓慢自转中的星系之内，尺度约为10万光年；它有若干条旋臂[①]，旋臂中的恒星绕着星系中心作轨道运动，大约每一亿年转过一周[②]。我们的太阳只不过是一颗中等大小的普通黄色恒星，它位于其中一条旋臂的外边缘。毫无疑问，自亚里士多德和托勒密以来我们经历了漫长的认识之路，而在他们那个年代地球被认为位于宇宙的中心。

① 旋臂：旋涡星系和棒旋星系中从星系核区或棒结构两端伸出的螺线形带状结构，主要由年轻亮星和星际介质构成。

② 太阳绕银河系中心转动一周约需两亿多年。

恒星的距离实在是太远了，以至于看上去它们只是一些非常小的光点。我们不可能确定恒星的大小和形状。那么，怎样才能把不同类型的恒星区分开来呢？对于绝大多

棒旋星系的旋臂

数恒星来说，唯一可以观测到，且不致发生误判的特征是它们的光的颜色。牛顿发现，如果使太阳光穿过一块棱镜，光线便会分解成构成阳光组成成分的各种颜色——太阳光谱，它看上去就像彩虹一样。类似地，把望远镜瞄准个别恒星或者星系并准确聚焦，就可以观测到恒星或星系的光谱。不同的恒星有不同的光谱，但不同颜色的相对亮度，总是会与某个灼热燃烧物体发出的光线所呈现的情况完全一样。这意味着可以由恒星的光谱来确定恒星的温度。还有，我们发现有一些特定的颜色在恒星光谱中是缺失的，而且这类缺失的颜色可以因恒星的不同而不同。我们知道，每一种化学元素都会吸收掉一组能表征有相应元素存在的特定的颜色。因此，只要把每一组这样的颜色与恒星光谱中缺失了的那些颜色相比对，就可以严格确认在恒星大气中存在有哪些元素。

牛顿 (1642—1727)

20世纪20年代，当天文学家开始观察河外星系中恒星的光谱时，异常情况发生了：它们所缺失的特征颜色组与我们的银河系中恒星的情况相同，但它们全都朝着光谱的红端移动，且相对位移量都一样。对此，唯一合理的解释是星系都在远离我们运动，因而星系光波的频率减小了，或者说发生了红移，其原因在于多普勒效应。请倾听一辆汽车在路上急驶而过的声音。当汽车由远方驶近时，汽车引擎声听起来音调比较高，相当于声波的频率比较高；当汽车由近处向远方驶离时，引擎声的音调听起来比较低。光波或辐射波具有类似的变化特性。实际上，警察正是利用多普勒效应，通过测定由汽车反射回来的无线电波脉冲的频率，来测出汽车的速度。

在证实了河外星系的存在之后，哈勃花了好多年时间来逐一记录星系的距离，同时还观测它们的光谱。在那个时候，大多数人都以为星系的运动是完全随机的，所以光谱呈现蓝移的星系应该与呈现红移的星系一样多。因此，当哈勃发现所有的星系都表现为有红移时，人们颇感意外，这说明每一个星系都在远离我们而去。更令人吃惊的是，哈勃在1929年发表的结果表明，甚至星系红移的大小也不是随机的，红移量居然与星系的距离成正比。换言之，星系越远，远离我们的速度就越快。因而，这意味着宇宙不可能如之前众人都猜想的那样是静态的，而是宇宙事实上正处于膨胀之中。在任何时刻，不同星系间的距离一直在不断地增大。

发现宇宙正在膨胀，乃是20世纪一项伟大的理性革命。事后来看，不禁让人惊讶为什么之前没有一个人想到这一点。牛顿等人应该会意识到，在引力的作用下一个静态宇宙很快会开始收缩。但是，请设想一下宇宙并不处于静止状态，而是正在膨胀。如果宇宙膨胀得不太快，那么引力的作用最终会使膨胀停止，并随之开始收缩。然而，要是膨胀速度超过某个确定的临界值，而引力作用不足以使膨胀停止，则宇宙便会一直不断地永远膨胀下去。这有点像我们在地球表面给火箭点火，使其上升时所发生的情况。如果火箭的速度比较慢，那么引力最终会使火箭停止运动，并随之开始向地面回落。要是火箭的速度大于某个临界值（约为每秒7英里①），引力便不足以把它拉回

① 即每秒11.2公里，也就是第二宇宙速度。

地面，于是火箭便会越飞越远，永远脱离地球。

在19世纪、18世纪，甚至17世纪晚期这段时间内的任何时候，都已经可以做到根据牛顿的引力理论来预言宇宙的上述变化特性。但是，人们关于静态宇宙的信念实在是太强了，这种信念一直延续到20世纪初。即使爱因斯坦在1915 年系统地阐明了广义相对论之时，他还是深信宇宙只能处于静止状态。因此，为了使静态宇宙成为可能，爱因斯坦对自己的理论做了修正，具体做法是在他的一些方程中引入了一个所谓的宇宙学常数①。这是一类新的"反引力"之力，与其他作用力的不同之处在于，这种力并非来自任何具体的力源，而是时空结构自身的组成部分。爱因斯坦的宇宙学常数给时空以某种固有的膨胀趋势，而且恰好可以与宇宙中全部物质的吸引力相平衡，这样一来自然会得出静态宇宙的结论。

① 宇宙学常数：爱因斯坦在建立他的"静止、有界、无限"宇宙模型时，人为引入的一个数值很小的常数。

看来，只有一个人愿意还广义相对论以其本来面目。尽管爱因斯坦和其他一些物理学家在不断探究各种途径，以能回避广义相对论所预言的非静态宇宙，俄国物理学家亚历山大•弗里德曼却与众不同地着手解释非静态宇宙。

弗里德曼模型②

② 弗里德曼模型：假设宇宙物质的大尺度分布为均匀各向同性的前提下所得到的一种最简单的膨胀宇宙模型，由苏联数学家弗里德曼于1922年得出。

广义相对论方程确定了宇宙如何随时间演化，然而这些方程的详细解算却极为复杂。因此，弗里德曼另辟蹊径，他就宇宙作了两个非常简单的假设：无论从哪一个方

亚历山大·弗里德曼
(1888—1925)

向去观察，宇宙看上去都是一样的；还有，要是我们能从任何别的地方观察宇宙，上述结论仍然成立。根据广义相对论和这两个假设，弗里德曼证明了我们不应该期望宇宙是静态的。实际上，在哈勃做出他的发现之前的若干年，弗里德曼于1922年就已精确预言了哈勃所发现的结果。

事实上，关于宇宙从任何方向看来都是相同的假设显然是不成立的。例如，我们银河系中的其他恒星在夜空中构成了一条明显的光带，这就是银河。但是，如果我们的观察对象是遥远的星系，那么从不同方向上看起来星系的数目大体上是相同的。所以，从不同方向去观察，宇宙看上去确实大体上是一样的，但这里有一个前提，即观测视野的范围应远远大于星系间的距离。

在很长的一段时间内，这为弗里德曼的假设——作为真实宇宙的某种粗略近似——提供了充足的理由。然而，后来一次很幸运的偶然事件揭示了真相：实际上弗里德曼的假设是对我们的宇宙的极为精确的描述。1965年，在新泽西州贝尔实验室工作的两位美国物理学家阿尔诺•彭齐亚斯和罗伯特•威尔逊，设计了一台甚高灵敏度的微波探测器，目的是用于与轨道上的卫星进行通讯联系。使两人深感迷惑不解的是，他们发现这台探测器所接收到的噪声比预期来得多，而且多余的噪声似乎并非来自任何特定的方向。开始时，他们寻找探测器上飞鸟的粪便，还检查了其他可能的仪器故障，但这些情况很快被一一排除。他们明白，对任何来自大气层内部的噪声来说，探测器倾斜安置时的噪声要比指向天顶时来得大，因为当探测器的指向与垂直方向成某个交角时，大气层会显得比较厚[①]。

无论探测器指向哪一方向，多余的噪声始终保持不变，所以它必然来自大气层之外。还有，尽管地球在不断地绕轴自转，同时又绕着太阳运动，但在整个一年中，无论白天还是黑夜，这种噪声始终保持不变。这说明辐射一定来自太阳系之外，甚至来自银河系之外，否则因探测器随地球运动而指向不同的方向，辐射也应当随之发生变化。

事实上，我们知道这类辐射在到达地球之前，必然穿越了可观测宇宙的大部分空间。因为辐射表现为各向同性，那么宇宙一定也是各向同性的，至少在大尺度上应该如此。现在我们知道，无论从哪个方向去看，这类噪声的相对变化绝不会超过万分之一。因此，彭齐亚斯和威尔逊在无意之中，以很高的精确度偶尔证实了弗里德曼的第一个假设。

差不多在同一时间，不远处普林斯顿大学的两位物理学家鲍勃·迪克和吉姆·皮伯尔斯也对微波饶有兴趣。当时他们正在深入研究乔治·伽莫夫的一种设想：早期宇宙应该是非常炽热的，且密度很高，会发出白热的光芒；须知伽莫夫曾经是弗里德曼的学生。迪克和皮伯尔斯认为，这种光芒现在仍然能看到，原因在于从早期宇宙非常遥远部分所发出的光线现在应当刚好到达地球。不过，由于宇宙膨胀，这种光线应该有非常大的红移，因而现在就

① 原文如此。实际含意是，这时沿探测器指向在大气层中所穿越的路径比较长，而不是大气层本身显得比较厚。

我们来看便表现为微波辐射。迪克和皮伯尔斯此时正在寻找这类辐射，当彭齐亚斯和威尔逊得知他们的工作时，便意识到自己已经找到了这种辐射。彭齐亚斯和威尔逊因这项工作于1978年获诺贝尔奖，而这对迪克和皮伯尔斯来说似乎有点残酷。

上述观测证据充分说明，无论在哪个方向上，宇宙看起来都是一样的，表面上看这好像暗示了我们在宇宙中所处的位置应该与众不同。说得具体一点，这似乎意味着如果我们观测到的所有河外星系都在远离我们而去，那么我们必然位于宇宙的中心。不过，对此也可以有另一种不同的解释：从任何其他的星系来看，在不同方向上所观测到的宇宙也许还是一样的。我们已经知道，这正是弗里德曼的第二个假设。

目前还没有任何科学证据来支持或者反对这个假设，我们只是谨慎地相信这一点。要是宇宙从我们周围的各个方向去看是各向同性的，而从宇宙中别的位置上去观察却并非如此，那就太不可思议了。在弗里德曼模型中，所有的星系都在彼此远离。这种情况有点像持续不断地吹一个表面上绘有若干斑点的气球。随着气球的膨胀，任何两个斑点之间的距离不断增大，但是任何一个斑点都不能被称为膨胀的中心。不仅如此，斑点间的距离越远，斑点之间互相远离的速度就越快。类似地，在弗里德曼模型中，任何两个星系之间互相远离的速度与星系间的距离成正比。所以，这个模型预言了星系的红移应该与星系的距离成正比，而哈勃所发现的恰恰就是这种情况。

尽管模型取得了成功，且预言了哈勃的观测结果，但弗里德曼的工作在西方一直鲜为人知。1935年，美国物理学家霍华德·罗伯逊和英国数学家亚瑟·沃克为说明哈勃发现宇宙均匀膨胀而提出了类似的模型，只是在这之后，弗里德曼的成就才为人们所知晓。

尽管弗里德曼只是发现了一个模型，事实上满足他的两个基本假设的却有三类不同的模型。在第一类模型，也就是弗里德曼所发现的模型中，宇宙膨胀得极为缓慢，以至于不同星系相互间的吸引力使得这种膨胀渐而减慢，并最终停止。之后，星系开始互相趋近，于是宇宙表现为收缩。相邻星系间的距离从零开始，不断增大到某个极大值，之后便逐渐互相接近，直到再次归复为零。

在第二类解中，宇宙膨胀得相当快，因此引力永远不可能使之停止，不过它会使膨胀速度稍有减慢。在这种模型中相邻星系间的距离从零开始，最终星系会以某种恒定的速度互相远离。

最后，还存在第三类解：宇宙膨胀速度的大小恰好能保证不会反转为坍缩。在这种情况下，星系间的距离还是从零开始并永远增大。然而，星系相互分离的速度会越来越慢，不过永远不会等于零[①]。

① 严格来说，在这类模型中星系向外膨胀的速度是以渐近的方式趋于零。

对第一类弗里德曼模型来说，一个值得注意的特征是宇宙在空间上并非是无限的，但也不存在任何边界。引力的作用之强使空间自行弯曲，情况犹如地球的表面。如果您在地球表面上沿着某个确定的方向一直不停地走下去，那么您永远不会遭遇到不可逾越的屏障，也绝不会从边缘处跌落下去，您最终会回到旅行开始时的出发点。在第一类弗里德曼模型中空间的情况也正如此，不过它是三维的，而不像地球表面只有二维。第四维——时间——在范围上也是有限的，但时间就像是一条线段，它有两个端点或说两个边界，即一个起点和一个终点。后面我们将会看到，只要把广义相对论与量子力学的测不准原理结合起来，就有可能做到空间和时间两者都是有限的，同时却没有边际或边界。您可以环绕着宇宙笔直地走下去，并最终会回到出发点——这一概念可衍生出绝妙的科幻小说题材，但这并没有多大的实际意义，因为可以证明在您还没有来得及兜上一圈时，宇宙的尺度早已重新坍缩为零了。要想在宇宙寿终正寝之前赶回起点，您的旅行速度必得超过光速，而这是不可能实现的。

但是，哪一类弗里德曼模型可用来描述我们的宇宙呢？宇宙最终会停止膨胀并随之开始收缩，抑或它会永远地膨胀下去？为了回答这个问题，我们需要知道两个数据：宇宙现在的膨胀速率和它目前的平均密度。如果这个密度小于某个确定的临界值，后者取决于膨胀速率，则吸引力就太弱而不足以使膨胀停止。要是密度大于该临界值，引力就会在未来某个时间使膨胀停止，宇宙会再度坍缩。

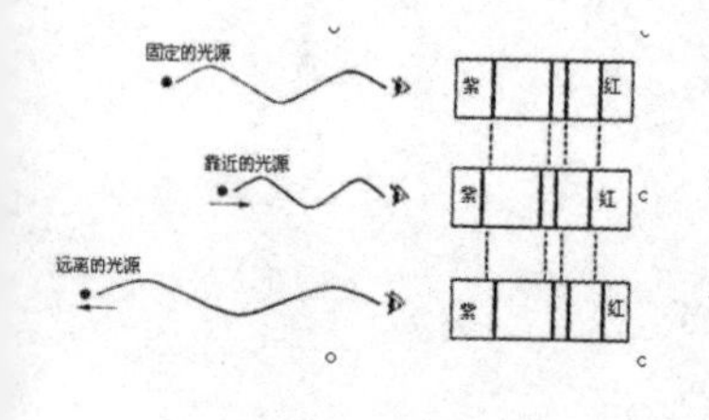

利用多普勒效应[①]，我们可以通过测量河外星系远离我们的运动速度，来确定宇宙目前的膨胀速度。这一步可以做得非常精确。然而，星系的距离只能通过间接的途径

① 多普勒效应：因声源（或辐射源）沿观测者视线方向运动，会引起观测者所接受到的声波（或辐射波）频率发生变化，称为多普勒效应，由奥地利物理学家多普勒于1842年首先发现。

来加以测定，测定结果并不太精确。因此，我们所能知道的宇宙膨胀速率也就是每10亿年在5%到10%之间。然而，我们对于宇宙目前平均密度的不确定性就更大了。

如果把我们在银河系和河外星系中所能观测到的全部恒星的质量相加，那么即使取膨胀速率的最低估值，恒星总质量还不到能使宇宙膨胀停止所需质量的百分之一。然而，我们知道在银河系和河外星系中必定包含了大量的暗物质①，尽管它们不可能被直接观测到，但鉴于暗物质的吸引力对星系中恒星和气体运动轨道的影响，我们可以确知它们必然存在。还有，大多数星系存在于星系团之中，我们可以通过类似的途径推知，在团内的星系之间应存在更多的暗物质，因为暗物质会影响到星系的运动。如果把所有这类暗物质加起来，其总质量仍只及能使膨胀停止所需质量的十分之一左右。尽管如此，也许还会存在尚未探测到的某种其他形式的物质，它们或许能使宇宙的平均密度增大到使膨胀得以停止所需的临界值。

① 暗物质：宇宙中不发射任何光和其他电磁辐射的物质，目前只能通过其引力效应来间接地加以探测。

据上所述，目前的证据意味着宇宙很可能会永远膨胀下去。但是，请勿对之深信不疑。我们真正可确认的全部事实是，即使宇宙将会再度坍缩，那也是遥远将来之事，至少得再过100亿年，因为宇宙至少已膨胀了这么长一段时间。对此我们无需过分担心，因为到那个时候除非我们已移民至太阳系之外，否则人类早已不复存在，早已随着我们的太阳的死寂而归于灭绝了。

大爆炸

所有弗里德曼解的一个共性特征是，在过去100亿至200亿年前的某一时候，相邻星系间的距离必然为零。这一时刻称为大爆炸，那时宇宙的密度和时空曲率[①]应均为无穷大。这意味着，作为弗里德曼解之基础的广义相对论预言了宇宙中存在一个奇点。

① 时空曲率：根据广义相对论，在引力场作用下时间和空间的性质取决于引力物质的分布，后者使时空发生弯曲，弯曲程度的大小可用时空曲率来表征，在大质量天体附近时空曲率是很大的。

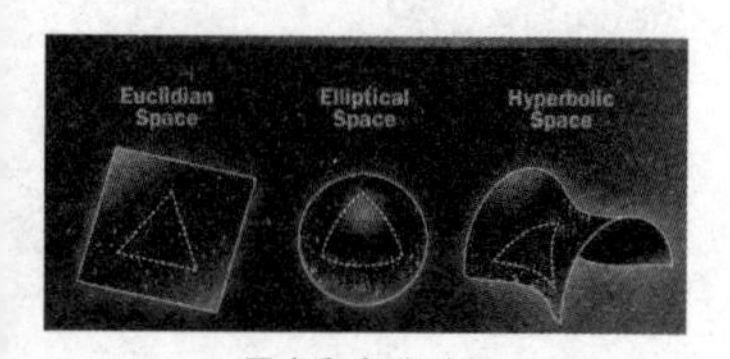

平直和弯曲时空

我们的全部科学理论体系之所以得以形成，乃是假设时空是光滑的，且近乎平直。因此，在大爆炸奇点处所有这些理论都不能成立，因为在那里时空的曲率为无穷大。这意味着即使在大爆炸之前确有事件发生，也不可能利用它们来推定其后会出现什么情况，原因在于在大爆炸时可预测性也会失效。由此可见，如果我们只知道大爆炸以来所出现的事，那就无法推定在大爆炸之前曾发生过些什么。就我们而言，大爆炸之前的事件是不可能产生任何效果的，因而这类事件不应成为科学宇宙模型的一部分。据此，我们应该把它们排除在模型之外，并宣称时间是有起点的，那就是始于大爆炸瞬间。

许多人不喜欢时间会有一个起点的观念，其原因可能在于这种观念有点像是掺入了神灵干预的味道。（另一方面，天主教会则充分利用了大爆炸模型，并于1951年正式宣称这一模型与《圣经》相一致。）为了回避有过一次大爆炸的结论，人们作了若干种尝试，其中得到最广泛支持的思想称为稳恒态理论。这一理论于1948年由三位学者共同提出，其中赫尔曼·邦迪和托马斯·戈尔德两人是来

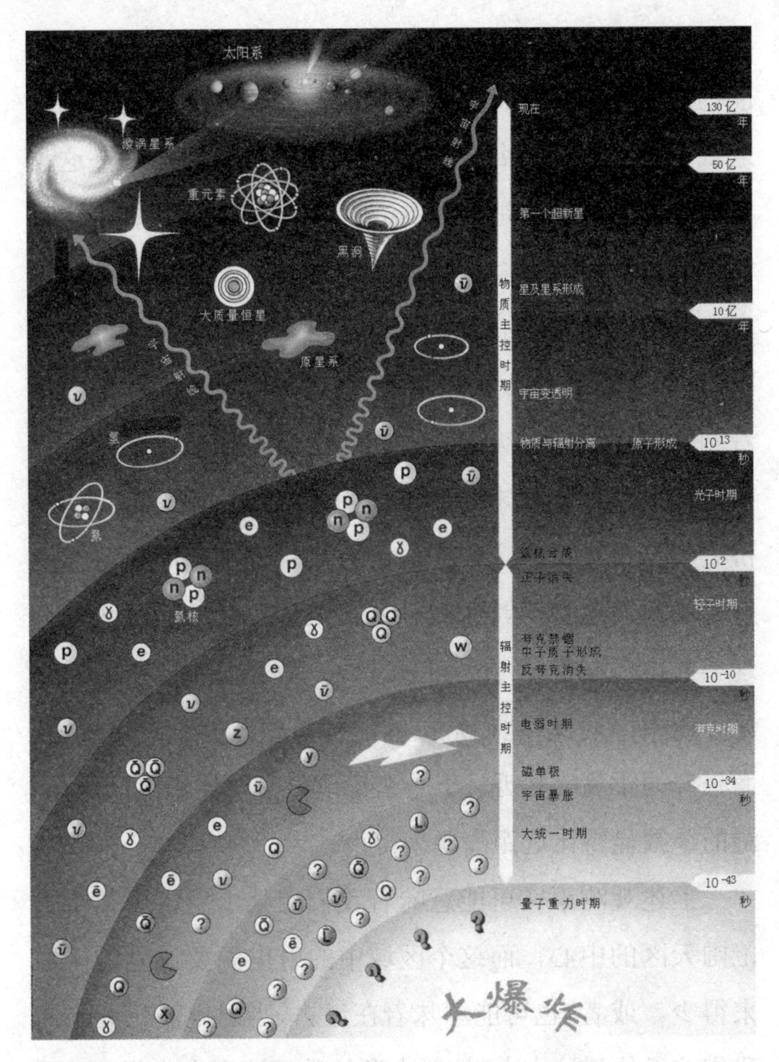

自纳粹占领下的奥地利的难民，另一位是英国人弗雷德·霍伊尔，后者与前两人一起从事战争期间的雷达研发工作。该理论的观念是，随着星系彼此间的互相远离，由于新的物质会连续不断地创生出来，一些新的星系便在原有星系之间的空隙中不断地形成。因此，不仅在空间的任何位置上，而且就不同的时间来看，宇宙的形态大体上都是相同的。

稳恒态理论要求对广义相对论加以某种修正，以保证物质能不断创生出来，不过所涉及的物质创生率非常之低，大约为每年、每立方公里创生出一个粒子——这与实验并不矛盾。这是一种不错的科学理论，优点在于它很简单，且能引出一些明确的、可通过观测来加以检验的预言。其中有一个预言是，无论从宇宙的哪个位置上来观察，也不管是在什么时间观察，任意的给定空间体积内所看到的星系或同一级天体的个数应该是相同的。

20世纪50年代末到60年代初，以马丁·赖尔为首的剑桥大学一批天文学家，完成了对来自外部空间射电波辐射源的巡天观测。剑桥大学这个小组的工作表明，大部分这类射电源必然位于银河系之外，而且弱源的个数比强源多得多。对此，他们给出的解释是弱源的距离比较远，而强源的距离比较近。于是在每单位空间体积内，近距离源的个数显得比远距离源来得少。

上述观测事实可能意味着我们应处于宇宙中某个大范围天区的中心，而这个区域中的射电源要比别的区域来得少。或者，也可能意味着在过去，当射电波仍处于向我们这里传播途中之时，射电源的数目比现在来得多。无论取哪一种解释，都与稳恒态理论所预期的结果相矛盾。再有，1965年彭齐亚斯和威尔逊所发现的微波辐射同样表明，宇宙过去的密度必然要高得多。因此，稳恒态理论不得不令人遗憾地被放弃。

为了回避必然有过一次大爆炸，因而时间必然有某个起点的结论，两位俄国科学家叶夫根尼·利夫希茨和伊萨克·哈拉特尼柯夫于1963年作了另一项尝试。他们提出大爆炸可能只是弗里德曼模型的一个特例，而这类模型充其量也不过是对真实宇宙的某种近似表述。也许，在所有能与真实宇宙大致相符的模型中，只有弗里德曼的模型才包含了一个大爆炸奇点。在弗里德曼模型中，所有星系之间只会沿径向运动并互相远离。这么一来，在过去的某个时间它们全都位于同一位置上也就不足为奇了。然而，真实宇宙中的星系并非严格按这种方式彼此远离，它们之间还会有少量的侧向速度[1]。所以，事实上根本无需要求全体星系在过去曾恰好位于相同的位置上，它们仅仅是彼此非常接近而已。因此，目前的膨胀宇宙也许并非始于大爆炸奇点，而是出现在更早期的某个收缩阶段[2]之后；随着那时宇宙的坍缩，宇宙中的粒子可能并没有全都碰在一起，粒子间也许只是交会而过，然后便互相远离，由此产生的结果正是现在看到的宇宙膨胀。那么，我们怎样才能得知真实宇宙是否确实始于一次大爆炸呢？

① 指与星系间连线（径向）相垂直的方向上的速度，天文学上称为切向速度。

② 收缩阶段，宇宙学中称为收缩相，下同。

利夫希茨和哈拉特尼柯夫所做的工作，是要研究这样一类宇宙模型，它们总体上与弗里德曼模型相类似，而同时又顾及真实宇宙中星系的不规则特性和随机运动。他们证明，即使星系不再始终保持沿径向彼此远离，这类模型仍可以一次大爆炸为起点。但是，他们认定这种情况只是在某些很特殊的模型中才有可能出现——模型中的所有星系必须全都按特定要求的方式运动。他们认为，可以提出两类弗里德曼模型，一类有大爆炸奇点，另一类则没有，但后者的个数比前者来得多，甚至多得不计其数，而由此应得出的结论是，出现过一次大爆炸的可能性实在非常之小。不过，他们后来又意识到，确有奇点存在的一般性弗里德曼类模型的个数还是很多的，而且模型中的星系也并非必须按某种特定的方式运动。据此，他们于1970年收回了自己提出的看法。

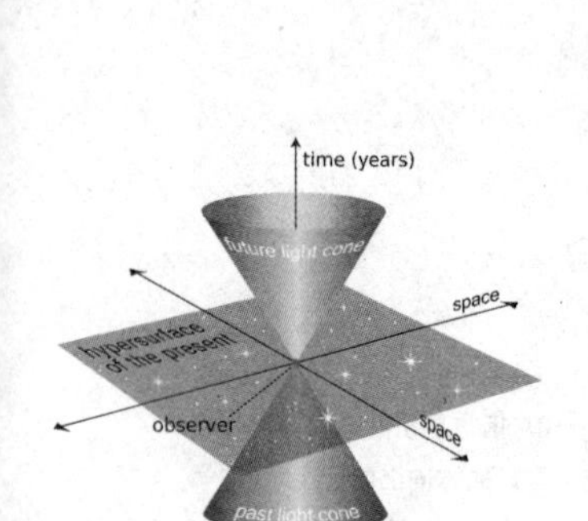

① 光锥：在理论时空中一束光随时间演化的轨迹，因其在三维空间中表现为一圆锥体而得名。

利夫希茨和哈拉特尼柯夫的工作是很有价值的，因为这项工作证明了，如果广义相对论是正确的话，那么宇宙可能有过一个奇点，即大爆炸。但是，它并没有解决一个关键问题：广义相对论能否预言我们的宇宙应该发生过大爆炸，即时间会否有起点？1965年，英国物理学家罗杰·彭罗斯开创性地通过另一条完全不同的途径为这个问题找到了答案。利用广义相对论中光锥[①]的变化特性，以及引力始终是吸引力这一事实，彭罗斯证明了在自引力作用下，处于坍缩中的一颗恒星必会落入某个区域之内，而该区域边界的尺度最终会收缩为零。这意味着该恒星中的全部物质将会收缩到一个体积为零的区域内，于是物质密度和时空曲率便变为无穷大。换言之，这就有了一个奇点，它位于被称为黑洞的时空区域之内。

表面上看，彭罗斯的结果完全没有涉及过去是否存在过大爆炸奇点的问题。不过，在彭罗斯得出他的定理之时，我还是一名研究生，并正在千方百计地寻找课题以完成我的博士论文。我意识到如果把彭罗斯定理中的时间方向倒过来，从而使坍缩变为某种膨胀，那么原理中的一些条件仍然可以成立，前提是目前所观测到的宇宙大尺度结构应大体上与弗里德曼模型相类似。彭罗斯定理已经表明，任何处于坍缩中的恒星必然终止于某个奇点；时间反演的论点指出，任何类弗里德曼膨胀宇宙必然始于一个奇点。鉴于一些技术上的理由，彭罗斯定理要求宇宙在空间上是无限的。因此，我就能利用这一原理证明，奇点应该存在的唯一条件是宇宙以足够快的速度膨胀，使它不会再次出现坍缩，因为唯有那种弗里德曼模型在空间上才是无限的。

在接下来的几年中，我推导出了一些新的数学方法，以从证明奇点必然会出现的那些定理中剔除这个以及其他技术性条件。最终结果见于彭罗斯和我本人联合发表的一篇论文，文中证明了必然存在过大爆炸奇点，前提条件只要求广义相对论是正确的，以及宇宙中所包含的物质与我们观测到的一样多。

对于我们的工作有不少反对意见，部分意见来自一些俄国学者，他们信奉由利夫希茨和哈拉特尼柯夫所奠定的思路，另一些持反对意见的人则感到凡涉及奇点的所有观念都是无法接受的，这会破坏爱因斯坦理论的完美形象。然而，人们毕竟不可能与数学原理争辩。因此，现在为人们广泛接受的观点是，宇宙必然有一个起点。

第3讲
黑洞

黑洞这一术语的出现乃是不久前的事。它是美国科学家约翰·惠勒在1969年创造出来的，用以形象化地描述至少可追溯到两百年前的一种观念。在那个时期，存在两类有关光的理论。一类理论认为，光是由粒子组成的，而另一类则主张光是一种波。现在我们知道，这两类理论实际上都是正确的。根据量子力学的波粒二象性，光既可以看作为波，也可以看作是粒子。就光是由波构成的理论而言，在引力作用下光会有何种表现是不清楚的。但是，如果光是由粒子组成的，那么就有可能对粒子在引力影响下的表现做出预言，而这时引力的作用方式与对炮弹、火箭以及行星是一样。

约翰·惠勒 (1911—2008)

在这一假设的基础上，剑桥的一位教师米歇尔于1783年在《伦敦皇家学会哲学学报》上发表了一篇论文。在该篇论文中米歇尔指出，如果一颗恒星的质量足够大，密度又足够高，那么恒星所具有的强引力场就有可能使光也无法逃逸掉。任何从恒星表面发出的光，还没有跑得太远就会在恒星引力的作用下被拽回来。米歇尔认为，这类恒星可能大量存在。鉴于它们所发出的光线不会到达我们这里，我们就不能看到这样的恒星；尽管如此，我们仍然能探测到它们的引力作用。这类恒星就是我们现在所说的黑洞，因为那是名副其实的——空间中的一些黑不可见的空洞。

几年以后，法国科学家拉普拉斯侯爵提出了类似的看法，而且他的工作显然与米歇尔无关。有意思的是，拉普拉斯的这一观念仅见之于他的专著《世界之体系》一书的第一和第二版，而在随后的各版本中再也没有出现；也许，拉普拉斯已认定这种观念太过荒唐。事实上，由于光速是恒定的，在牛顿引力理论中像对炮弹那样来处理光就必然会出现矛盾。由于引力的作用，从地球上向上发射的炮弹会渐而减速，最终便告停止，并随之落回地面。但是，光子会以恒定的速度持续不断地向上运动。那么，牛顿引力能以何种方式影响到光呢？直到爱因斯坦于1915年提出广义相对论之时，一种关于引力如何影响光的自洽理论才得以问世；而且即便如此，只是在又过了很长一段时间之后，人们才真正明白了爱因斯坦理论对大质量恒星的含意。

为了理解黑洞是怎样形成的过程，首先需要弄清楚恒星的生命周期。当大量的气体（其中大部分是氢）在自引力的作用下开始坍缩，最终便会形成一颗恒星。随着气体的收缩，气体中原子间的碰撞变得越来越频繁，同时运动速度越来越大，其结果是气体的温度不断升高。最终，气体的温度变得非常之高，以致氢原子间不再因碰撞而相互弹开，而是会并合在一起形成氦原子。这种反应犹如受控氢弹，而反应所释放的热量就是使恒星闪闪发光的原因。由此产生的热量还会使气体的压力增大，直到压力足以与引力相平衡时气体便不再收缩。这种情况有点像气球内部空气的压力与气球胶皮张力之间的平衡：空气压力力图使气球膨胀，而胶皮张力则力图使气球缩小。恒星会在很长的一段时间内维持这样的稳定状态，即核反应产生的热量与引力相平衡。然而，恒星最终会把内部的氢和其他核燃料消耗殆尽。而且，恒星形成之初所含有的核燃料越多，它把燃料耗尽所花的时间就越短，这看上去有点不合常理。原因在于，恒星的质量越大，能与引力取得平衡所需的温度就越高，而温度越高，燃料消耗的速度便越快。对我们太阳来说，所含有的燃料很可能足以再用上50亿年左右，但更大质量的恒星可以在1亿年这么短的时间内把燃料耗尽，这要比宇宙年龄小多了。一旦燃料耗尽，恒星便会冷却下来，于是它就开始收缩。之后又可能发生什么情况，对此最早的认识已经是20世纪20年代的事了。

1928年，一位名叫苏布拉马尼扬•昌德拉塞卡的印度研究生乘船赴英格兰，拟就读于剑桥，并师从英国天文学家亚瑟•爱丁顿爵士。爱丁顿是一位广义相对论的行家。这里有一则故事，说是有一位旅行家曾于20世纪20年代初询问爱丁顿，他听闻世界上仅有三个人理解广义相对论。爱丁顿对此的回答是："我正想知道这第三个人究竟是谁。"

苏布拉马尼扬·昌德拉塞卡
(1910—1995)

在从印度出发的这次旅行途中，昌德拉塞卡完成了一项工作：质量多大的恒星能在全部燃料消耗殆尽后，仍

沃尔夫冈·泡利 (1900—1958)

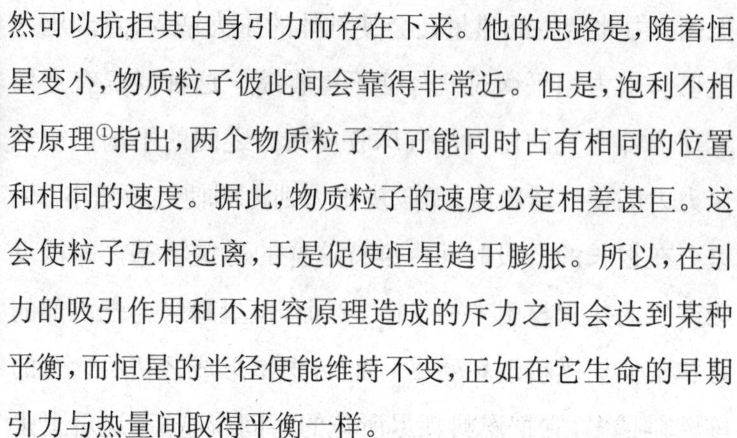

然可以抗拒其自身引力而存在下来。他的思路是，随着恒星变小，物质粒子彼此间会靠得非常近。但是，泡利不相容原理[1]指出，两个物质粒子不可能同时占有相同的位置和相同的速度。据此，物质粒子的速度必定相差甚巨。这会使粒子互相远离，于是促使恒星趋于膨胀。所以，在引力的吸引作用和不相容原理造成的斥力之间会达到某种平衡，而恒星的半径便能维持不变，正如在它生命的早期引力与热量间取得平衡一样。

① 泡利不相容原理：原子中不可能容纳运动状态完全相同的两个或两个以上的电子，由奥地利物理学家泡利于1925年提出。

然而，昌德拉塞卡意识到，对不相容原理所能提供的斥力来说，存在某一个限值。相对论限制了恒星中物质粒子运动速度的最大差异不得超过光速。这意味着当恒星密度变得足够高时，不相容原理引起的斥力应当小于引力的吸引作用。昌德拉塞卡的计算表明，对于一颗无能源的恒星来说，当它的质量大于约1.5倍的太阳质量时，这颗恒星便不可能抵抗其自引力的作用而维持现状不变。现在，人们把这个质量称为昌德拉塞卡极限。

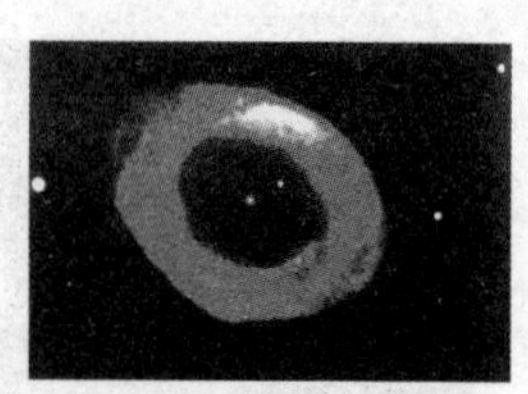

中间是一颗白矮星

这一点对大质量恒星的终极归宿有着极为重要的意义。如果质量小于昌德拉塞卡极限，恒星最终会停止收缩，并安然进入一种可能的终极状态，成为一颗白矮星[2]，半径为几千英里，密度达到每立方英寸数百吨。白矮星就是由恒星物质中电子间的不相容原理斥力来维持的。我们已观测到了大量的这类白矮星。第一个被发现的白矮星是绕着天狼星运动的那颗恒星，而天狼星是夜空中最明亮的恒星。

② 白矮星：恒星演化末期所形成的一种致密星，体积与地球相近，密度可高达10^5—10^7g cm^{-3}，质量不超过太阳质量的1.44倍。

人们又意识到，对于一颗质量范围约为一至两倍太阳质量的恒星来说，还存在另一种可能的终极状态，但其尺度甚至比白矮星还要小得多。维持这类恒星的力，应当来自中子和质子（而不是电子）间的不相容原理斥力。正因为如此，它们便称为中子星[①]。中子星的半径只有10英里左右，而密度则达到每立方英寸数亿吨。当人们首次对中子星做出预言之时[②]，还没有任何方法可以观测到中子星，探测到中子星已是好多年之后的事了。

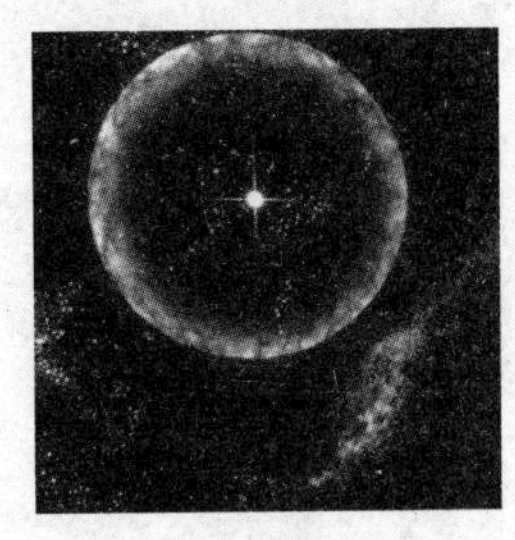
中子星

① 中子星：大质量恒星演化末期所形成的一种致密星，因主要成分为中子而得名；半径约10公里，密度高达10^{14}—10^{15}g cm^{-3}，质量上限为2—3倍太阳质量。

② 1932年苏联物理学家L.D.朗道首先从理论上预言有可能存在中子星，1939年美国科学家奥本海默等建立了第一个中子星模型。

另一方面，对质量超过昌德拉塞卡极限的恒星来说，当它们走到燃料耗尽这一步时会出现很大的问题。在一些情况中，恒星可以发生爆炸，或者通过某种方式抛去足够多的物质，这样一来它们的质量便会低于昌德拉塞卡极限，然而要确信无论恒星有多大总会发生这类事件是很困难的。如何才能知道恒星必定会损失质量？而即使每一颗恒星都会通过某种途径失去足够多的质量，那么要是对白矮星或中子星补充更多的质量使之超过昌德拉塞卡极限，又会出现何种情况？恒星是否会持续坍缩下去，直至密度达到无穷大呢？

爱丁顿对此感到震惊，他拒不接受昌德拉塞卡的结论。爱丁顿认为，恒星绝无可能会坍缩成一个点。这也正是大多数科学家的观点。爱因斯坦本人发表过一篇文章，他断言恒星不会收缩为零尺度。其他一些科学家也对此持反对意见，特别是爱丁顿，须知爱丁顿曾是昌德拉塞卡的导师，又是关于恒星结构研究方面的最大权威，而这些意见便促使昌德拉塞卡放弃了他的工作思路，并转而从事天文学其他问题的探索。然而，1983年昌德拉塞卡被授予诺贝尔奖，这至少有一部分是鉴于对他有关无能源恒星极限质量之早期研究工作的肯定。

昌德拉塞卡已经证明，对一颗质量大于昌德拉塞卡极限的恒星来说，不相容原理不可能使其坍缩过程停止下来。但是，如何依据广义相对论来推测这类恒星会发生些什么情况的问题，则一直要到1939年才由一位年轻的美国人罗伯特•奥本海默给出解答。不过，他的结论表明，借助当时的望远镜不可能探测到任何观测结果。后来，二次大战不期爆发，奥本海默本人全身心地投入到了原子弹计划之中。战后，有关引力坍缩的问题基本上已被人遗忘了，因为那时大多数科学家的兴趣已专注于原子和原子核尺度上所发生的现象。然而，在20世纪60年代，随着现代技术用于天文观测，观测对象的数量和范围大大地

爱因斯坦与奥本海默

扩大了，从而重新激活了人们对天文学和宇宙学中一些大尺度问题的兴趣。这时，一些学者再度注意到了奥本海默的工作，并对之加以发展。

根据奥本海默的工作，我们现在可以得到如下的图像：与无恒星存在的情况相比，由于恒星引力场的存在，会使光线在时空中的路径发生改变。光在发出后，它在时空中行进的路径可以用光锥来表述，光锥的顶点即为发出瞬间光所处的位置，而光锥会朝向恒星表面略有弯曲。这种现象在日全食时是可以观测到的，表现为来自远方恒星的星光出现了偏折。随着恒星的收缩，恒星表面的引力场越来越强，而光锥向内弯曲的程度亦渐而显著。在这一过程中，光要逸出恒星便变得越来越困难，而且对远方的观测者来说，星光会逐渐变得更暗、更红。

最终，当恒星收缩到某个确定的临界半径时，恒星表面引力场会变得非常强，其结果是光锥向内弯曲的程度之大使光再也不可能从恒星向外逸出。根据相对论，任何物体的运动速度都不可能大于光速。因此，如果光无法逸出，那么任何其他物体也就不可能向外逸出。这么一来，就会存在一个事件集合，或者说一个时空区域，任何事物都不可能从该区域逸出而到达远方的观测者。我们现在把这个区域称为黑洞，黑洞的边界称为事件视界，事件视界与刚好不能从黑洞逸出的光线路径相一致。

如果您正在注视一颗恒星坍缩为黑洞的过程，那么为了理解您会看到的情况，必须牢记在相对论中是不存在绝对时间的。每个观测者都有自己的时间量度。对于某颗恒星上的一个人来说，他的时间与远方另一个人的时间是不相同的，原因在于恒星有引力场。这一效应已经在地球上所做的实验中，通过安放在水塔顶端和底部的计时钟测出来了。设想在一颗坍缩中恒星的表面有一位无所畏惧的宇航员，他根据自己的表，每隔1秒钟向绕着这颗恒星运转的空间飞船发出一个讯号。在他表上的某个时间，比如说11点，恒星收缩到了临界半径之内，这时引力场变得非常强，以至于讯号再也不可能到达他的飞船了。

对于留在飞船上观察的伙伴们来说，他们应当发现随着11点的不断逼近，那位宇航员所发出的一个接一个讯号间的时间间隔会变得越来越长。不过，在10时59分59秒之前，这种效应是很不明显的。在宇航员的10时59分58秒讯号与宇航员的表为10时59分59秒时所发出的讯号之间，伙伴们所必须等待的时间仅比1秒略为长了一点点，然而若要想收到11时的讯号，他们必须得无限期地永远等下去。根据宇航员的表，光波是从10时59分59秒与11时之间从恒星表面发出的，而从飞船上来看，那光波将绵延于无穷大的时间间隔里。

在飞船上，依次到达的光波之间的时间间隔会变得越来越长，因而星光会显得越来越红，也越来越暗。最后，恒星会变得非常之暗，而从飞船上就再也不能看到它了。这时，所剩下的就只是空间中的一个黑洞。不过，恒星仍然会对飞船施以相同的引力作用。这是因为对飞船来说恒星依然是可观察的，至少原则上应该如此。只是由于恒星引力场的作用，恒星表面发出的光有非常大的红移，结果便不可能看到了。但是，红移不会影响到恒星自身的引力场。因此，飞船仍会绕着黑洞继续作轨道运动。

彭罗斯和我在1965至1970年间所做的一项工作证明，根据广义相对论，在黑洞内部必然存在着一个密度无穷大的奇点。情况有点像时间起点时的大爆炸，但对坍缩中的天体和那位宇航员来说，这应当是时间的终点。在奇点处，科学定律以及我们预测未来的能力一概失效。不过，这种预测能力的失效并不会影响到留在黑洞外的任何观测者，因为无论是光，还是其他什么讯号，都不可能到达他们那里。

这个引人注目的事实导致彭罗斯提出宇宙监督假设，这一假设从含义上也许可理解为"上帝嫌弃裸奇点"。换句话说，由引力坍缩造成的奇点只能出现在像黑洞那样的地方，奇点在那里被事件视界严严实实地隐藏了起来，外部观测者根本就看不到。严格说来，这正是所谓的弱宇宙监督假设：它保护了留在黑洞外的观测者，奇点处出现的预测能力失效的种种后果对其是没有影响的。但是，对不幸落入黑洞的那位可怜的宇航员来说，这一假设却无任何的保护作用。难道上帝不也应该保护他的体面吗？

在广义相对论方程的某些解中，有可能使我们的那位宇航员看到裸奇点。他也许能做到避免与奇点相遇，而是落入并穿过一个"虫洞"①，出现在宇宙的另一个区域中。这一结果应当为时空旅行提供了一些绝妙的可能性。

① 虫洞：宇宙中可能存在的、连接两个不同时空区域的狭窄隧道，由奥地利物理学家弗莱姆于1916年首次提出。

虫洞示意图

然而，遗憾的是所有这些解似乎都是非常不稳定的。极小的一点扰动，譬如一名宇航员的存在，就有可能使这类解发生变化，从而使这位宇航员在碰上奇点并到达其时间终点之前，不可能看到这个奇点。换言之，奇点永远处于宇航员的未来，绝不会出现在他的过去。

宇宙监督假设的强版本指出，在一个现实的解中，奇点要么就像引力坍缩中的奇点那样永远出现在未来之中，要么便会像大爆炸那样完全见之于过去，二者必居其一。人们渴望某种版本的宇宙监督假设能得以成立，因为在接近裸奇点的地方也许有可能实现到过去时代去旅行。尽管对科幻小说家来说这应当是一个妙不可言的题材，但一旦付诸实现，任何人的生命将不再会永远安全。有人也许会回到过去，在您尚未成为胎儿之前就把您的父亲或者母亲杀死了。

在引力坍缩并形成黑洞的过程中，运动会被引力波的发射所阻断。因此，可预料到的情况是，无需太长时间黑洞便会平静下来，并处于某种稳恒状态。过去人们通常认为，这种终极稳恒状态应当取决于经坍缩而形成黑洞的那个天体的具体细节。黑洞可能大小不一，形状各异，而且它们的形状甚至有可能不是固定不变的，而是在不停地脉动。

沃纳·伊斯雷尔(1931—)

然而，1967年沃纳·伊斯雷尔在都柏林发表的一篇论文使关于黑洞的研究发生了革命性的变化。伊斯雷尔证明了任何无自转的黑洞，必然呈现完美的圆球形。不仅如此，黑洞的大小应当由质量唯一地确定。实际上，这可以用爱因斯坦方程的一个特解来表述，这个特解是在广义相对论面世后不久的1917年由卡尔•史瓦西得出的。一开始，伊斯雷尔的这一结果，被包括他本人在内的许多人解释为黑洞只能从具有完美圆球形的天体坍缩而成的证据。鉴于任何一个真实的天体都不会是完美无缺的圆球体，上述结论意味着一般情况下引力坍缩会导致形成裸奇点。但是，罗杰•彭罗斯和约翰•惠勒对伊斯雷尔的结果给出了另一种解释，而且解释得非常细致。他们认为，黑洞的行为应该像一个液体球。尽管一个天体的初始状态并非为圆球形，但随着它的坍缩并形成黑洞，由于引力波的发射，这个天体会平静下来，并最终成为圆球状态。后来更详细的计算支持了这种观点，并最终为人们所普遍接受。

伊斯雷尔的结果只涉及到由无自转天体形成的黑洞这一种情况。与液体球相类似，人们会想到由一个有自转的非完美圆球形天体所形成的黑洞。由于自转的效应，这样的黑洞在其赤道周围应当表现出某种隆起。我们在太阳上观测到了因自转引起的这类隆起，而太阳的自转周期约为25天[1]。1963年，新西兰人罗伊•克尔发现了一组广义相对论的黑洞解，而且比史瓦西解更具有普遍性意义。这类“克尔”黑洞以恒定的速率自转，其大小和形状只取决于黑洞的质量和自转速率。如自转速率为零，黑洞便具有完美的圆球形，这时的克尔解与史瓦西解完全一致。但是，如果自转速率不为零，黑洞便会在其赤道附近向外隆起。因此，人们自然会推测，对于一个有自转的天体来说，

① 太阳是一个气体球，自转情况与固体地球不同，太阳赤道附近转得最快。

它经历坍缩过程而形成黑洞的终极状态应当用克尔解来描述。

1970年，我的一位同事和研究生同学布兰登·卡特为证实这一推测迈出了第一步。他指出，如果一个以恒定速率自转的黑洞像一个自转的陀螺一样，有一个对称轴，那么黑洞的大小和形状应当只与它的质量和自转速率有关。之后，我于1971年证实，任何以恒定速率自转的黑洞确实应当具有这样一个对称轴。最后，到了1973年，伦敦国王学院的戴维·鲁宾逊利用卡特和我的结果证明，上述推测是正确的：这类黑洞确实必然是克尔解。

因此，黑洞经引力坍缩后一定会平静下来，它可以有自转，但并不出现脉动式变化。还有，黑洞的大小和形状应当只取决于它的质量和自转速率，与经坍缩而形成黑洞的那个天体的性质无关。这一结果被戏称为"黑洞无毛"。这意味着天体经坍缩而形成黑洞后，有关这一天体的许多信息全都丢失掉了，因为之后对这个天体有可能加以测定的全部信息仅限于它的质量和自转速率。在下一讲中我们将会看到这一结论的重要意义。无毛定理对黑洞的可能类型做出了严格的限制，所以它也具有非常重要的实际意义。因此，人们可以对有可能包含黑洞的那些天体构筑一些详细的模型，并把这些模型的预测与观测结果加以比较。

从科学史上看，一种理论先借助数学模型进行非常详细的推导，之后才通过观测取得证据以说明它的正确性，这种情况为数并不很多，而黑洞可算是其中一例。实际上，一些对黑洞持反对意见的学者就曾经把这一点作为他们的主要理由。须知，有关这些天体的唯一证据是根据广义相对论计算出来的，而这种理论又未必完全靠得住，那么人们怎样才能相信它们呢？

然而，在1963年，加利福尼亚帕洛马山天文台的一位天文学家马尔滕•施密特发现了一个暗弱的恒星状天体，该天体位于名为射电波源3C273的方向上，这里3C273指的是剑桥第三射电源表中编号为273的射电源。他测得了该天体的红移，结果发现其红移惊人之大，因而不可能是由引力场造成的：如果这是一种引力红移，那么这个天体必然具有极大的质量，而且应该离我们非常近，以至于会影响到太阳系中行星的运动轨道。由此说明红移必另有起因，即起因于宇宙膨胀，而这又意味着该天体的距离非常遥远。既然在这么大的距离上还能看到它，这个天体必须非常亮，而且它所发出的能量必然大得出奇。

为了找到能产生如此大能量的原因，唯一可取的机制就是引力坍缩，且不是一颗恒星的坍缩，而是星系整个中央区域的坍缩。之后，又陆续发现了若干类似的其他"类恒星状天体"，即类星体，而且它们全都有很大的红移。不过，所有这些类星体都非常遥远，很难借助观测来提供黑洞存在的决定性证据。

有关黑洞存在的后继进展出现于1967年，剑桥的一位研究生乔丝琳•贝尔发现，天空中有一些天体在不断地发出很有规则的脉冲式射电波。最初，乔丝琳和她的导师安东尼•休依什以为，也许他们接触到了银河系中的某类外星文明。我确实还记得，在宣布这项发现的一次讨论会上，他们把第一批发现的四个源命名为LGM 1—4，LGM代表"小绿人"①。

① "小绿人"(Little Green Men)是国外童话故事中的一种外星人形象。

但是后来，他们以及所有其他的学者终于得出了一个不太含有浪漫色彩的结论：这类天体事实上只是一些自转中子星，并被命名为脉冲星。由于中子星的磁场与其周围物质有着复杂的间接性相互关系，它们会不断发出脉冲射电波。这样的结果使描述空间探索的那些西部小说的作者深感不快，然而对当时我们中间相信有黑洞存在的少数人来说却是非常鼓舞人心的。这是关于存在中子星的第一项明确无误的证据。中子星的半径约为10英里，只有恒星转变为黑洞时所需临界半径的几倍大。如果一颗恒星可以坍缩到如此小的尺度，那么由此推想其他一些恒星有可能坍缩到更小尺度并成为黑洞就未必没有道理了。

既然根据自身明确的定义，黑洞不会发出任何光辐射，那么又怎样才有望能探测到它们呢？这似乎有点像在煤窖里寻找一只黑猫。幸好，对此还是有办法的——正如约翰·米歇尔在1783年他的那篇开创性论文中所指出的那样，黑洞的引力仍然会对邻近天体产生影响。天文学家已经观测到一些天体系统，其中的两颗恒星因彼此间的引力吸引而做互绕运动。他们也发现，在某些这类系统中只能看到一颗恒星，它绕着某个不可见的伴星做轨道运动。

当然，不能马上由此得出伴星就是一个黑洞的结论，它也许只是一颗因为太暗而没有看到的恒星。然而，某些这类系统同时还是强X射线源，如天鹅X-1即为其中一例。对这类现象的最合理解释是，X射线是由可见恒星表面抛出的物质产生的。随着抛出物向不可见伴星下落，它展现出某种旋涡式运动——就像水从浴缸中流出来一样，而且变得非常炽热，发出X射线。为使这种机制得以发挥作用，不可见天体必须非常小，如白矮星、中子星，或者黑洞。

现在，从观测到的可见恒星的运动状况，可以确定不可见天体的最小质量。就天鹅X-1来说，这一质量约为太阳质量的6倍。根据昌德拉塞卡的结果，如果不可见天体是一颗白矮星，那么这个数字就太大了。因此，看来它必然是一个黑洞。

还有其他一些模型可用来解释天鹅X-1而无需涉及黑洞，不过它们全都显得相当牵强附会。看来，黑洞是对观测结果唯一最为自然的解释。尽管如此，我曾与加州理工学院的基普•托恩打过赌：天鹅X-1实际上并不含有黑洞。对我而言，这是某种形式的保险策略。关于黑洞我做了大量的工作，如果最终发现黑洞并不存在，那么所有这些工作便全都白费劲了。但是一旦出现这种情况，我将会得到些许安慰，那就是赢得我的赌注，即获赠四年的《私家侦探》杂志。要是黑洞确实存在，那么基普将只能获得为期一年的《阁楼》杂志，因为当我们在1975年打这个赌时，已有80%的把握知道天鹅座的那个天体是一个黑洞。现在，我要说的是这种可能性已达到95%左右，不过这笔赌债已经结清了。

有证据表明，在我们的银河系内的其他一些天体系统中有黑洞存在，而且在河外星系和类星体的中心还存在质量大得多的黑洞。我们还可以考虑这样一种可能性，即也许会存在一些比太阳质量小得多的黑洞。这类黑洞不可能通过引力坍缩过程形成，因为它们的质量小于昌德拉塞卡极限。对于这类小质量恒星来说，即使在内部核燃料耗尽之时，它们仍能自行维持与引力间的平衡。因此，能形成小质量黑洞的唯一条件是，在非常大的外部压力的作用下，物质能被压缩到具有极高的密度。这种条件有可能出现在非常大的氢弹中。物理学家惠勒曾经做过一项计算：如果能把地球上全部海洋中所有的重水[①]都提炼出来，就有可能制成一枚氢弹，而这枚氢弹会把中心区的物质高度压缩，并最终生成一个黑洞。不过遗憾的是，

① 重水：由重氢（氘）和氧组成的化合物，分子式为D_2O，因分子量比普通水（H_2O）略高而称为重水，在天然水中重水含量约占0.015%。

黑洞示意图

那时没有人能存活下来去观察它了。

一种更为现实的可能性是，这类小质量黑洞也许已经在极早期宇宙的高温、高压条件下形成了。如果早期宇宙的物质分布并非绝对平滑和完全均匀，那么就有可能形成黑洞，原因在于那时某个小区域的物质密度会高于平均密度，它会通过上述方式经压缩而形成黑洞。然而，我们知道过去必定存在过一些密度分布不规则区，不然的话今天宇宙中物质的分布应该仍然保持完全均匀的状态，而不会集聚成恒星和星系了。

为了说明恒星和星系的存在需要有密度的不规则分布，而这种不规则性会不会导致相当大数目的此类原初黑洞的形成，则取决于早期宇宙中诸多条件的细节。所以，要是我们能确定目前所存在的原初黑洞的个数，就应当会获得有关宇宙极早期阶段的许多认识。对于质量大于10亿吨（相当于一座大山的质量）的原初黑洞来说，只能通过它们的引力作用对其他可见物质，或者对宇宙膨胀的影响来加以探测。然而，我们在下一讲中将会明白，黑洞并非完全黑不可知：它们会像灼热物体那样发出辐射，而且黑洞的质量越小，所发出的辐射越强。由此可见，与大黑洞相比，较小质量的黑洞实际上也许更容易探测到，这听起来显得有点不合常理。

第4讲
黑洞并非黑不可知

1970年之前，我个人关于广义相对论的研究，主要集中于是否存在过大爆炸奇点的问题上。然而，那一年11月份我女儿露西出生后不久的一个晚上，我在就寝之际开始思考起黑洞来。由于我的残疾人体质，思考过程颇为缓慢，因而花了我好多时间。在那个年代，关于时空中有哪些点位于某个黑洞之内，又有哪些点位于黑洞之外，尚无明确的定义。

当时我已和彭罗斯讨论过给黑洞下一个定义的想法，即把黑洞定义为事件的某种集合，而这些事件不可能逸出一段大的距离。现在，这正是人们所普遍采用的定义。它意味着黑洞的边界，或者说事件视界，是由恰好无法摆脱黑洞的那些光线构成的。这些光线永远会在黑洞的边缘止步不前。情况就像一个人在摆脱警察的追捕，他始终能保持跑得快一步，但却不能彻底逃脱掉。

突然间我意识到，这些光线的路径是不可能彼此趋近的，因为一旦彼此趋近，它们必然最终会碰在一起。这有点像另有一人沿着相反方向在逃离警察。结果是两个人会一起被逮着，或者说在这种情况下都跌入了黑洞。但是，要是这些光线被黑洞所吞噬，那么它们便不可能曾一度出现在黑洞的边界上。所以,事件视界处的光线必然彼此沿平行方向运动,或者是互相远离。

理解上述图像的另一条途径是,事件视界(即黑洞边界)就好比是阴影的边缘。它是光线逸出一段大的距离之边缘,但同样也是即将到来的厄运之阴影的边缘。如果观察一个远距离光源(如太阳)所投出的阴影,你将会看到边缘处的光线是不会彼此趋近的。要是来自事件视界(即黑洞边界)的光线永远不会互相趋近,那么事件视界的面积就可以保持不变,或者随时间而增大。这一范围永远不会变小,因为变小就意味着至少有一部分边界上的光线必然彼此趋近。事实上,一旦有物质或辐射落入黑洞,事件视界的面积就会增大。

再有,设想有两个黑洞发生了碰撞且并合在一起,形成单一的一个黑洞。这时,所形成的黑洞之事件视界的面积,会大于两个原始黑洞的事件视界面积之和。事件视界面积这种永不变小的性质,对黑洞可能存在的行为给出了一个重要的限制。我为我的这一发现兴奋不已,以至于那天晚上睡得不太好。

第二天我给罗杰•彭罗斯打了电话。他对我的看法表示赞同。我认为,实际上之前他已经认识到了事件视界面积的这种性质。不过,他一直采用的关于黑洞的定义稍有不同。他还没有意识到的是,只要黑洞已经不再活动并处于某种稳恒状态,那么根据这两种定义所推知的黑洞边界应当是一样的。

热力学第二定律[①]

黑洞面积这种永不变小的行为，使人马上联想到有一种物理量的变化特性，那就是熵，它可以用来量度一个系统的无序程度。常识告诉我们，如果没有外来因素的影响，系统的无序程度总是在增大的；只需对一座不加修理的房子观察其变化就会明白这一点了。我们可以从无序中创造出有序——例如，可以对房子进行粉刷。不过，这样做需要消耗能量，因而可资利用的有序能量的数量也就减少了。

能对上述概念给出精确描述的乃是热力学第二定律。该定律指出，一个孤立系统的熵永远不会随时间而减小。不仅如此，要是两个系统合而为一，那么合并后系统的熵会大于单个系统的熵之和。例如，试考虑一只盒子内的气体系统。我们可以把分子想象为一些微型桌球，它们会不断地互相碰撞，也会从盒子的壁上弹回来。假定在最初时刻，用一块隔板把所有的分子都限制在盒子的左半部。然后，一旦把隔板抽掉，这些分子一定会扩散开来，并充满盒子的左右两半部。在之后的某个时刻，它们会非常偶然地全部进入盒子的右半部，或者全部都回到左半部。但是，在绝大部分时间内，盒子两半部中的分子数目总是大致相等的。这种状态与全部分子都位于半只盒子内的初始状态相比，有序程度较低，而无序程度较高。于是，我们就说气体的熵增大了。

① 热力学第二定律：热力学的基本定律之一，有多种表达方式，是指如没有外界的影响，系统的温度只能由高到低变化，一切与热运动有关的物理、化学过程都具有不可逆性。

类似地，可以设想开始时有两只盒子，其中一只盒子内装的是氧分子，另一只装的是氮分子。如果把这两只盒子连接起来，再把中间的壁去掉，这时氧分子和氮分子便开始互相混合。在之后的某个时刻，最可能出现的状态是，在两只盒子的所有地方，氧分子和氮分子会完全均匀地混合在一起。与两只盒子分开时的初始状态相比，现在这种状态的有序程度较低，因此熵就比较大。

热力学第二定律所表述的内容，与别的一些科学定律颇为不同。其他定律，譬如以牛顿引力定律为例，这是一种绝对定律，也就是说它们始终是成立的。与之相反，热力学第二定律是一种统计定律，即它并不能永远成立，而只是在绝大多数情况下可以成立。在后来的某个时刻，盒子中全部气体分子出现在某半只盒子中的概率远小于万亿分之一，不过这种情况毕竟还是有可能出现的。

但是，如果附近有一个黑洞，要想破坏热力学第二定律就显得不那么难了：你所要做的只是，把某种熵很大的物质（比如一盒气体）抛向黑洞。这么一来，黑洞之外物质的总熵就会减小。当然，你仍然可以说包括黑洞内部的熵在内的总熵并没有减小。不过，既然我们无论如何也没法观察到黑洞的内部，也就不可能知道黑洞内部物质的熵有多大。因此，如果黑洞存在某些特征，而位于黑洞外的观测者可以利用这些特征来探知黑洞的熵，那情况就很妙了；一旦有含熵的物质落入黑洞，黑洞的熵应该会增大。

我发现只要有物质落入一个黑洞，事件视界的面积就会增大，而普林斯顿大学一位名为雅科布·贝肯斯坦的研究生便根据这一发现提出，事件视界的面积可以用来量度黑洞熵的大小。随着含熵物质落入黑洞，事件视界的面积应当会增大，于是黑洞外物质的熵与黑洞视界面积之和就永远不会减小了。

这一想法在大多数情况下看来能避免违反热力学第二定律。不过，这里存在一个致命的缺陷：要是黑洞有熵，那么它也应该有温度。但是，一个有非零温度的物体，必定会以某种确定的速率发出辐射。常识告诉我们，如果你把一支火钳放入火中加热，火钳便会变得又红又热，并发出辐射。然而，低温物体也是会发出辐射的，通常情况下人们只是没有注意到这一点而已，原因在于辐射量非常之低。为了不致违反第二定律，这种辐射必须予以考虑。所以，黑洞应当会发出辐射，但根据其自身的定义，黑洞应该是不会发射出任何东西的物体。可见，黑洞事件视界的面积好像也不能被视为就是它的熵。

事实上，1972年我曾就这一议题与卡特以及一位美国同事吉姆·巴丁合作写过一篇论文。当时我们曾指出，尽管熵与事件视界的面积有不少相似之处，但却存在这样一个显然是致命的难点。我必须承认，促使我写这篇论文的部分原因是受到贝肯斯坦工作的刺激，因为我感到他错用了我关于事件视界面积增大的发现。但是，最终发现，他基本上还是正确的，不过他肯定没有料想到正确之处在哪里。

黑洞辐射

1973年9月，在访问莫斯科期间，我与两位权威苏联专家雅科夫·泽利多维奇和亚历山大·斯塔罗宾斯基讨论了有关黑洞的问题。他们使我确信，根据量子力学的测不准原理，自转黑洞应该会产生并发射粒子。我对他们在物理学基础上所作的一些推论深信不疑，但并不喜欢他们在计算粒子发射时所用的数学模式。于是，我着手设计了一种更好的数学处理方法，这种方法我曾于1973年11月底在牛津举办的一次非正式研讨会上做过介绍。当时，我并没有通过具体的计算来确认实际上会发射出多少粒子。我曾期望所能发现的辐射，恰好就是泽利多维奇和斯塔罗宾斯基从自转黑洞所预测到的结果。然而，经过计算后我发现，即使是无自转的黑洞，它们显然也应该会以某种恒定的速率产生并发射粒子，这令我惊讶不已，并深感迷惑不解。

一开始我以为，这种辐射表明计算时我所采用的若干项近似中，有一项是不成立的。我担心如果贝肯斯坦发现了这一点，他会用来作为又一个论据，以支持其关于黑洞熵的观念，而对此我仍然并不喜欢。但是，对这一问题的思考越是深入，我越是感到事实上那些近似是应当成立的。不过，最终使我相信这种辐射确实存在的事实是，被辐射出去粒子的谱恰好正是热物体的辐射谱。黑洞在以恰到好处的速率不断地发射出粒子，从而保证不致违背第二定律。

从那时起，其他人又通过若干种不同的形式反复进行了计算。他们全都证实了黑洞应当像有温度的热物体那样会发出粒子和辐射，而这里的温度仅取决于黑洞的质量：质量越大，温度越低。可以通过以下方式来理解这种发射：被我们设想为真空的空间不可能完全空无一物，不然的话各种场，如引力场和电磁场等，都必然严格为零。然而，场的强度及其随时间的变化率可类比为粒子的位置和速度。根据测不准原理，对其中的一个量知道得越精确，另一个量就越不可能测准。

所以在虚无空间中，场是不可能始终保持严格为零的，不然就会出现场的强度值恰好为零，而同时它的变化率也恰好为零。实际情况是，就一个场的强度而言，必然存在某种最小的不确定性量值，或者说量子起伏。我们可以把这种起伏设想为光或引力的粒子对，它们在某个时刻同时出现，因运动而彼此远离，然后再度相遇并互相湮没。这类粒子称为虚粒子。虚粒子与实粒子不同，它们不可能直接用粒子探测器来加以观测。不过，它们的一些间接效应，如电子轨道和原子的能量之微小变化则是可以测出来的，而且以异乎寻常的精确度与理论预期值相吻合。

根据能量守恒，虚粒子对中的一个成员会具有正能量，而另一个成员则有负能量。负能量成员必然是一种短命的粒子，这是因为在通常情况下实粒子总是具有正能量。因此，负能量粒子必须找到它的伙伴，并与之湮灭。然而，黑洞内部的引力场非常之强，即使是实粒子也可以具有负能量。

所以，如果有黑洞存在，那么具有负能量的虚粒子有可能落入黑洞，并变为实粒子。在这种情况下，这个虚粒子不再必须与它的伙伴发生湮灭；它那被遗弃了的伙伴同样有可能落入黑洞。不过，因为它具有正能量，也可能作为一个正粒子而逃逸至无穷远。对一定距离外的一名观测者来说，这个粒子便表现为是由黑洞发射出来的。黑洞越小，负能量粒子在变为实粒子之前所必须越过的距离就越短。随之而来的是发射率便越大，而黑洞的表观温度就越高。

向外辐射的正能量会与落入黑洞的负能量粒子流取得平衡。根据爱因斯坦的著名方程式$E=mc^2$[①]，能量与质量是相当的。因此，由于负粒子流落入黑洞，黑洞的质量就会减小。随着黑洞质量的损失，黑洞事件视界的面积便逐渐减小，但是黑洞熵的这种减小会因所发出辐射的熵得以补偿，而且是超额的补偿，可见这绝没有违反热力学第二定律。

① $E=mc^2$：质能转换公式，表示物质的质量和能量可以互相转换，揭示了能量和质量间的内在联系，由爱因斯坦于1905年提出。

黑洞爆炸

黑洞的质量越小，它的温度就越高。所以，随着黑洞质量的损失，它的温度和发射率便逐渐增高。于是，黑洞质量的损失就变得更快。当黑洞质量最终变得极小之际将会出现何种情况，我们对此还没有非常清晰的认识。最合理的推测是，黑洞会通过一次爆发式的终极发射而完全消失，其辐射能量之大可相当于数百万颗氢弹的爆炸。

对于一个质量为太阳的若干倍的黑洞来说，温度应当仅为绝对温标[1]的千万分之一度。这比宇宙中无处不在的微波辐射的温度低得多，后者约为绝对温标2.7度——所以这类黑洞释放出的能量应小于它们所吸收的能量，尽管后者也是非常之小。如果宇宙命中注定要一直不断地永远膨胀下去，那么微波辐射的温度最终会减小到低于这类黑洞的温度。那时，黑洞所吸收的能量将会小于发射出去的能量。不过，即使到了那个时候，黑洞的温度仍然非常之低，要完全蒸发殆尽大约会需要10^{66}年。这个数字远比宇宙的年龄长得多，后者仅约为10^{10}年。

另一方面，我们在上一讲中已经知道，也许还存在质量极小的原初黑洞，它们是在宇宙的极早期阶段中，由不规则密度分布区因坍缩而形成的。这类黑洞应当有高得多的温度，发出辐射的速率也会大得多。对于一个初始质量为10亿吨的黑洞来说，它的寿命大体上与宇宙的年龄相等。初始质量更小的黑洞，应当已经完全蒸发掉了。然而，质量稍大一些的黑洞现在仍然会以X射线和伽马射线[2]的形式在发出辐射。这些射线与光波相类似，不过波长要短得多。这类黑洞很难称得上是黑的。它们实际上是白热的，正以约为1万兆瓦的功率发射能量。

要是我们能驾驭这样一个黑洞的能量输出，那么它可以抵得上十座大型发电站。不过，要想做到这一点相当困难。这个黑洞的质量相当于一座大山，却被压缩成原子核般大小。如果地球表面上有一个这样的黑洞，那么它会

① 绝对温标：亦称开氏温标，用符号K表示。是建立在卡诺循环基础上的热力学温标。规定摄氏零度以下273.15℃为零点，称为绝对零点。

② X射线和伽马射线：波长范围在0.1—10纳米之间的电磁辐射形式称为X射线，而波长比X射线更短的电磁辐射称为伽马射线。

洞穿地面并向地球中心落去，任何方法都不能使它停下来。这个黑洞会穿透地球来回振荡，直到最终在地心处安居下来。所以，想要有可能利用它所发射的能量，安置这样一个黑洞的唯一地点是应当把它放在环绕地球的轨道上。而且，可以使它绕地球作轨道运动的唯一途径是，在它的前方拖动一个大质量物体，以把黑洞吸引到那里去，这种情况有点像在驴子面前放上一根胡萝卜。这种设想听起来不太现实，至少在近期内无法实现。

搜索原初黑洞

但是，即使我们不可能利用这些原初黑洞所发出的辐射，我们又是否有机会能观测到它们呢？我们可以寻找原初黑洞在大部分生存时间内所发出的伽马射线。大多数原初黑洞的距离都很遥远，它们所发出的辐射非常微弱；尽管如此，全部此类黑洞的总体效应也许是可以探测到的。事实上，我们确实观测到了这类伽马射线背景。不过，产生这种背景辐射的过程可能并不起因于原初黑洞。我们可以说的是，伽马射线背景的观测结果并没有为原初黑洞提供任何确凿的证据。但是，这些观测结果告诉我们，平均来说宇宙中每立方光年内，这种微黑洞的数目不可能超过300个。这一上限意味着，原初黑洞充其量也只能占到宇宙平均质量密度的百万分之一。

既然原初黑洞乃罕见之物，那么看来也许不可能会有一个这样的黑洞，其距离之近使得我们可对这一个体进行观测。但是，鉴于任何物质的引力作用会使原初黑洞集聚起来，在星系内它们的数密度就应该大得多了。比如说，要是星系内原初黑洞的数密度大上100万倍，那么离我们最近的黑洞就有可能位于大约10亿公里远的地方，与已知的最远行星冥王星的距离差不多[①]。在这样的距离上，即使黑洞稳定发射的功率达到1万兆瓦，想要探测到它仍然是很困难的。

① 原文如此。冥王星现已被重新归类为矮行星，它到太阳的平均距离约为60亿公里。

为了观测到一个原初黑洞，必须在合理的时间段（例如一星期）内，在同一方向上检测出几个伽马射线量子。不然的话，所检测到的量子也许只不过是伽马射线背景的一部分。不过，普朗克的量子原理[②]告诉我们，每个伽马射线量子都有很高的能量，这是因为伽马射线所处的频段非常高。因此，即使辐射功率高达1万兆瓦，也无需太多的量子。为了观测到来自冥王星那么远的地方的为数不多的量子，所需要的伽马射线探测器应当比迄今已建成的任何同类探测器都来得大。还有，鉴于伽马射线无法穿透大气层，探测器一定要安置在太空中。

马克斯·普朗克 (1858—1947)

② 普朗克量子原理：光和其他电磁波只能发射或吸收分立的量子，而量子的能量与其频率成正比；该理论由德国物理学家普朗克于1900年首先提出，故又称普朗克假设。

当然，要是一个黑洞距离很近，处于冥王星的位置上，且已到达其寿命的结束期并发生爆炸，那么检测它的爆发式终极发射就不是一件难事了。然而，如果这个黑洞在过去的100亿至200亿年内一直不断地在发出辐射，那么它会在接下来的若干年内到达其寿命结束期的可能性实际上是相当小的。在过去或将来的几百万年内，也许同样有可能出现过、或者将出现此类事件。所以，为了在您的研究经费用完之前抓着能观测到一次黑洞爆炸的合理机会，您必须找到一种方法，以能检测到大约1光年距离范围内的任何爆炸事件。还有一个问题是，您需要一台大型伽马射线探测器，以能观测到由爆炸产生的几个伽马射线量子。不过，这种情况下已没有必要去确认来自同一方向的所有量子。现在要做的只是观测在非常短的时间间隔内所到达的全部量子，并能合理认定这些量子来自同一次爆发。

切伦科夫 (1904—1990)

有一种伽马射线探测器也许有能力找出原初黑洞，那就是地球的整个大气层（我们无论如何不太可能有能力来建成比这更大的探测器）。一旦有一个高能伽马射线量子击中地球大气中层的原子，它就会产生出一些正负电子对。当这些电子对又击中其他一些原子时，便会继而产生更多的正负电子对。这样一来便出现了所谓电子簇射的现象，其结果是产生某种形式的光，称为切伦科夫辐射[①]。因此，通过对夜空中光闪烁的搜寻，便可能探测到伽马射线暴[②]。

① 切伦科夫辐射：高速荷电粒子在介质中的运动速度大于介质中光速时所产生的一种特殊辐射，具有明显的方向性和强偏振等特性，1934年由苏联物理学家切伦科夫首先发现。

② 伽马射线暴：天体伽马射线辐射在短时间内突然极度增强的现象，典型持续时间为数秒钟，通常认为起因于超新星爆发，后者称为伽马射线暴源。

当然，还存在若干种其他现象（如闪电），它们也能造成天空中的闪光。但是，伽马射线暴与这一类效应是可以区分开来的，办法是在距离相隔很远的两个或两个以上的地方同步观测闪光现象。来自都柏林的两位科学家尼尔•波特和特雷弗•威克斯，已经利用位于亚里桑那的望远镜做过一项此类探索。他们发现了若干次闪光，但都不能确认为是原初黑洞引起的伽马射线暴。

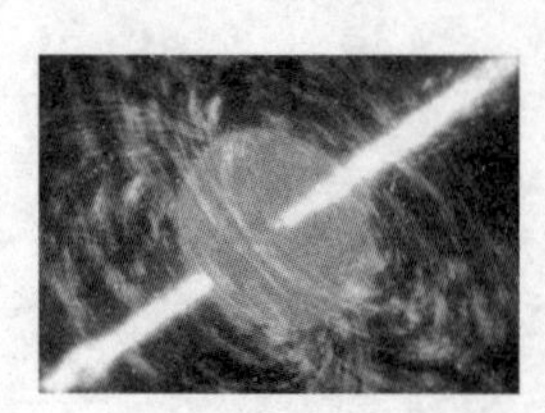

伽马射线暴示意图

现在看来对原初黑洞搜索的结果可能确是否定的，但即便如此，这一结果仍然会给我们提供有关宇宙极早期阶段的重要信息。如果早期宇宙始终处于混沌或不规则状态，或者说如果那一时期物质的压力一直很低，那么可预期的情况是，这种状态所产生的原初黑洞，应当比由我们的伽马射线背景观测所确定的黑洞的限值要多得多。只有当早期宇宙非常平滑、均匀，而且压力又高时，才能解释可观测到的原初黑洞数目之缺损现象。

广义相对论和量子力学

黑洞辐射乃是依据20世纪两个伟大理论——广义相对论和量子力学——所共同预言的第一个例子。起初它引来了一片反对声，因为它颠覆了既有的观点：“黑洞怎么能发射出什么东西来？”当我在牛津附近卢瑟福实验室召开的一次会议上第一次宣布我的计算结果时，我受到了普遍的质疑。在我的报告结束之时，会议主席，来自伦敦国王学院的约翰•泰勒就断言我的报告纯属无稽之谈。他甚至还为此写过一篇论文。

不过，最后包括泰勒在内的大多数人终于接受了这样的结论：如果我们有关广义相对论和量子力学的其他观念都是正确的话，那么黑洞必定会像热物体那样发出辐射。因此，尽管我们还没能设法找到一个原初黑洞，但比较一致的共识是，如果找到了，黑洞必然在发出大量的伽马射线和X射线。如果我们确实找到了一个，我将会获得诺贝尔奖。

我们曾经认为，引力坍缩是不可逆转的终极过程，而看来黑洞辐射的存在便意味着这种观念不再成立。如果一名宇航员跌入了某个黑洞，那么黑洞的质量就会增大。最终，与增加部分质量相当的能量会以辐射的形式返回宇宙。因此，从某种意义上说，这位宇航员将得以重生。然而，这类重生而不朽并无多大意义，因为当宇航员在黑洞内部粉身碎骨而不复存在时，他个人的任何时间概念几乎肯定已走到尽头。甚至由黑洞最终发射出的粒子，一般来说在类型上也会与构成宇航员的粒子相迥异。对于这位宇航员而言，他得以留存下来的唯一特征应当是他的质量或能量。

我在推导黑洞发射的过程中曾用到了一些近似算法，而当黑洞质量大于若干分之一克时这些近似颇为有效。但是，在黑洞寿命结束之际，当黑洞质量变得非常小时，这种近似就失效了。看来，最有可能的结局是黑洞会恰好消失，至少是从我们这一区域的宇宙中消失不见。随之消失的有那位宇航员，以及在黑洞内部也许存在的任何奇点。这可算是第一个迹象，说明量子力学有可能回避经典广义相对论所预言的奇点。不过，我和其他人在1974年用于研究引力之量子效应的一些方法，并不能回答诸如奇点会否在量子引力中出现这样一类的问题。

因此，从1975年起我开始推求一种更为有效的途径，以根据费因曼对历史求和的思想来研究量子引力。沿着这条途径可以就宇宙的起源和归宿问题给出一些回答，这将在下面的两次讲座中予以说明。我们将会看到，量子力学允许宇宙有一个非奇点的开端，这意味着在宇宙诞生时无需要求物理学定律失效。宇宙的状态及其所包含种种内容，包括我们自身在内，在达到测不准原理设定的极限之前，完全由物理学定律所确定。自由意志的空间仅此而已。

第5讲
宇宙的起源与归宿

在整个20世纪70年代，我的主要工作一直是研究黑洞。然而，1981年当我参加在梵蒂冈举行的一次宇宙学讨论会时，我对涉及宇宙起源的一些问题的兴趣再度被激起。当天主教会试图为一个科学问题立法，宣称太阳在绕地球转动时，曾对伽利略犯下了一个极为恶劣的错误。① 几个世纪后的今天，教会认定了比较好的做法是应当邀请一批专家就宇宙学方面为它提供建议。

① 伽利略于1632年出版的《两种世界体系的对话》中，得出了关于地球围绕太阳运转的证明，被教会指控为违背了教皇的命令，并推广与《圣经》相左的学说。1633年罗马宗教法庭对伽利略的审判，是历史上科学与宗教冲突最为激烈的事件之一。

在会议结束时，与会者获准谒见教皇。他告诉我们，研究大爆炸之后的宇宙演化并无不当，但不应该探究大爆炸本身，因为此乃创生时刻，故而应为上帝之杰作。

当时，令我欣慰的是教皇并不知晓我刚在会上所作报告的题目。我可不想重蹈伽利略命运之覆辙；我对伽利略寄以很大的同情，原因之一在于我恰好是在伽利略逝世300周年那一天出生的。

加利莱奥·伽利略
(1564—1642)

热大爆炸模型

为了说明我的那篇论文所谈及的内容，首先我将根据被称为“热大爆炸模型”的学说，来阐述人们所普遍接受的宇宙演化史。这一学说承认，自大爆炸以来，宇宙可由弗里德曼模型表述。在这类模型中您会发现，随着宇宙的膨胀，宇宙中物质和辐射的温度在不断下降。因为温度就是对粒子平均能量的一种量度，这种冷却过程便会对宇宙中的物质施以重大的影响。在温度非常高的时候，粒子会以极高的速度朝着不同的方向运动，结果是粒子不可能因核力和电磁力的吸引作用而彼此集聚在一起。但是，随着温度的降低，可预料到的情况是粒子会互相吸引并开始聚集起来。

在大爆炸瞬间，宇宙的尺度为零，因而温度必然为无穷大。但是，随着宇宙的膨胀，辐射的温度会不断下降。在大爆炸之后的1秒钟，温度会降低到约100亿度。这大约是太阳中心温度的1000倍，不过氢弹爆炸时就会达到这么高的温度。在这一时刻，宇宙的主要成分应当是光子、电子、中微子[①]以及它们的反粒子，同时还会有一些质子和中子。

① 中微子：静止质量为零、与物质的作用最弱、且以光速运动的中性基本粒子。

随着宇宙继续膨胀，温度进一步下降，在碰撞过程中电子和电子对的产生率，会变得低于它们因湮灭而消失的速率。于是，大部分电子和反电子会彼此湮灭，产生出更多的光子，只剩下为数不多的电子。

大约在大爆炸后的100秒，温度会下降到10亿度，这也是最灼热恒星内部的温度。到达这一温度时，质子和中子所具有的能量已不足以摆脱强核力的吸引作用。它们开始可以结合在一起，生成氘（即重氢）原子核，其中包含了一个质子和一个中子。然后，氘核又会与别的质子和中子结合，生成含有两个质子和两个中子的氦核。此外还会生成少量的两种较重的元素，即锂和铍。

可以计算出，在热大爆炸模型中，大约有四分之一的质子和中子会转化成氦核，同时还生成少量的重氢和其他一些元素。多余的中子衰变为质子，也就是普通氢原子的核。这些理论预期值与观测结果非常吻合。

热大爆炸模型还预言，我们应该能观测到从早期灼热阶段所遗留下来的辐射。不过，由于宇宙膨胀，这种辐射的温度应当已降低到绝对温标几度。这就解释了彭齐亚斯和威尔逊在1965年所发现的微波背景辐射。因此，我们完全确信已取得了正确的图像，至少可以追溯到大爆炸后的一秒钟左右。在大爆炸后仅仅过了几个小时，氦和其他元素的产生过程即告停止。而且，在这之后接下来的约100万年时间内，宇宙只是表现为继续膨胀，而没有发生太多的其他事情。最终，一旦温度跌至几千度时，电子和原子核便不再具有足够的能量来克服它们之间电磁力的吸引作用。这时，它们就会开始结合在一起，并生成原子。

从整体上看，宇宙仍然会继续膨胀，同时温度继续降低。但是，在那些密度略高于平均密度的区域内，额外的引力吸引作用会使膨胀减慢下来。这一过程最终会使某些区域不再继续膨胀，并再次出现坍缩。在坍缩过程中，由于区域之外物质的引力作用，这些区域就有可能开始呈现少量的自转。随着坍缩区范围渐而变小，自转速度会越来越快——这种情况就像在冰上做旋转动作的滑冰者，一旦他们把双臂收紧，转动的速度就会加快。最后，当这类区域变得足够小时，其转动速度之快足以与引力作用取得平衡。有自转的盘状星系就是通过这种方式诞生的。

随着时间的推移，星系中的气体会碎裂成一些较小的云块，它们会在自身引力的作用下发生坍缩。收缩过程中气体的温度会增高，一旦温度变得足够高时，核反应就开始了。这类反应又会使氢转变为氦，期间所释放出的热量使压力增大，于是云块不再进一步收缩。这种状态的云块便是像我们的太阳那样的恒星，它们可以维持很长的时间，期间氢燃烧转变为氦，所产生的能量则以热和光的形式向外辐射。

超新星爆发

对质量更大的恒星来说，由于引力作用更强，需要有更高的温度与之取得平衡。于是，核聚变反应会进行得非常之快，在大约只有1亿年的时间内恒星的氢燃料便会消耗殆尽。这时，它们会表现为略有收缩，并随着温度的进一步升高开始把氦转变为更重的元素，如碳和氧。然而，这一过程不会释放出太多的能量，于是危机便出现了，那就是我在有关黑洞的那一讲中所描述的场景。

一颗新星

人们还没有完全弄清楚接下来将会发生些什么情况，不过看来恒星的中心区有可能会坍缩成某种非常致密的状态，如中子星或者黑洞。恒星可能在一次剧烈的爆炸中把它的外层向外抛出，这就是超新星[①]爆发，此时恒星的亮度会超过星系中所有其他的恒星。恒星在行将寿终正寝之际所产生的一些较重元素，会被抛回到星系内

① 超新星：光变幅度最大的一类变星，恒星亮度会在短时间内增强千万倍甚至上亿倍，甚至在白天也能看到。新星：爆发变星的一种，亮度可在几天内突然增大几万倍，然后在几个月甚至更长的时间内渐而减暗至原有的亮度。

的气体中，它们为生成下一代恒星提供了部分原材料。

我们自己的太阳含有2%左右此类较重的元素，因为它是一颗第二代（或第三代）恒星。太阳在大约50亿年前由一块自转气体云形成，而气体中含有更早时期超新星的碎屑。云块中的大部分气体经演化而形成太阳，或者被向外吹走。然而，有少量较重的元素会聚集在一起，并形成绕太阳做轨道运动的天体——行星，地球即是其中之一。

尚未解决的问题

宇宙从最初的极端高温状态开始，并随膨胀而冷却的图像，与今天我们所取得的所有观测证据都是一致的。尽管如此，仍有几个重要的问题尚未得以解决。首先，为什么早期宇宙会有如此高的温度？其次，为什么宇宙在大尺度上会如此均匀——为什么在空间中的不同位置、以及从不同的方向上看宇宙都是一样的？

第三，为什么宇宙最初的膨胀速率会如此接近临界值，从而恰好保证不会再度坍缩？如果大爆炸后一秒钟时的膨胀速率哪怕只是小了10亿亿分之一，宇宙就会在达到它今天的大小之前再度坍缩。另一方面，要是一秒钟时的膨胀速率增加同样的数值，那么宇宙就会极度膨胀，以至于现在它简直就会变得空无一物了。

第四，尽管事实上宇宙在大尺度上表现为高度均匀和各向同性，但其中不乏存在局部性的物质团块，如恒星和星系。人们认为,这些天体是因早期宇宙中不同区域内存在少量密度差异而演化形成的。试问,这类密度涨落的起因是什么?

仅仅依据广义相对论不可能解释这些特征，或者说无法对这些问题给出解答。这是因为广义相对论预言,宇宙最初时的密度为无穷大,也就是始于大爆炸奇点。在奇点处,广义相对论和其他所有的物理学定律全都失效。我们不可能预言从奇点会发展出什么样的东西来。正如前面我已解释过的那样，这意味着理论上同样可以不考虑大爆炸之前发生的任何事件，因为这类事件对我们来说是没有任何观测效应的。时空应当有一个边界,亦即发端于大爆炸。宇宙为什么应该从大爆炸瞬间开始,以一种确定的方式演化,并最终成为我们今天所观测到的状态呢?为什么宇宙会如此均匀，而且恰到好处地以临界速率膨胀,从而不致发生再一次坍缩呢?如果能够证明,有着多种不同初始结构的宇宙，都会演化成我们今天所观测到那种状态,那么人们便应当更为高兴了。要是情况确实如此,那么从某类随机性初始条件发展而来的宇宙,应该包含了若干个我们今天所观测到的那种区域。也许还会存在一些与之很不相同的区域，不过这类区域可能并不适合于星系和恒星的形成。星系和恒星是进化成智慧生命所必须具备的重要先决条件,至少就我们所知应该如此。因此，这些区域就不会包含能观测到它们不同之处的任何生命。

在研究宇宙学问题时,必须考虑到选择原理,即我们生活在宇宙中一个适合智慧生命的区域中。这个显而易见的基本因素有时候被称为人择原理[1]。相反，试想宇宙的初始状态只有在经过极为仔细的选择后，才能保证会演化出我们在自己周围所看到的那些事物。如是，那么宇宙就不大可能包含任何会出现生命的区域。

[1] 人择原理：该原理认为，正是因为有人类存在，才能解释宇宙的种种特性；如宇宙不具有现在的形态，就不会有人类这样的智慧生命来认识宇宙。

在前面我已介绍过的热大爆炸模型中，早期宇宙阶段并没有足够的时间能使热量从一个区域传递到另一个区域。这意味着在诞生之初，宇宙中的不同区域必定有着严格相同的温度，只有这样才能说明下列事实：我们所看到的微波背景在不同方向上有着相同的温度。还有，宇宙膨胀的初始速率必然经过非常精确的选定，从而保证在今天之前宇宙不会再次坍缩。这就意味着，如果热大爆炸模型自时间起点以来都是正确的话，那么宇宙的初始状态确实作了非常仔细的选择。要想解释宇宙为什么恰好应该以这种方式诞生是很困难的，除非借助上帝之手——上帝的本意就是要创造出我们这样的生命。

暴胀模型

为了避免热大爆炸模型在极早期阶段的上述困难，麻省理工学院的艾伦·古思提出了一种新的模型。在他的模型中，许多不同的初始结构都可以演化成如目前宇宙的那种状态。他认为，对早期宇宙来说，可能在一段时间内作极高速的指数式膨胀。这种膨胀称为“暴胀”——类似于每个国家中在一定程度上都会出现的物价暴涨。物价暴涨的世界纪录也许当推第一次世界大战后的德国，当时一只面包的价格从原来的不到一马克，在几个月时间内涨到数百万马克。不过，在宇宙尺度上可能出现过的暴胀甚至比这还要大得多，仅仅在一秒钟的极小一部分时间内，宇宙就膨胀了100万亿亿亿倍。当然，那时尚未有现在这样的政府。

古思认为，宇宙从大爆炸诞生之际温度极高。可以预料，在这样高的温度下，强核力、弱核力和电磁力全都会统一成单一的一种力。宇宙的温度会随膨胀而降低，同时粒子的能量应随之减小。最后，应当出现所谓相变①，而力与力之间的对称性便会发生破缺。强力会变得与弱力和电磁力有所不同。一个常见的相变例子就是把水冷却使其结冰。液态水是对称的，在不同的位置或者不同的方向上都没有差异。但是，一旦冰晶体形成后，这些晶体会有着确定的位置，而且会沿着某一方向排列成行。这么一来就破坏了水的对称性。

就水而言，如果处理得当，可以使它处于“过冷”状态。这就是说，可以把水的温度降到冰点（0摄氏度）以下，但不会结冰。古思的观点是，宇宙的特性可能会以类似的方式发生变化：温度有可能跌至临界值之下，而力与力之间的对称性却并没有出现破缺。要是发生了这种情况，那么宇宙便会处于某种非稳定态，此时的能量会比发生对称性破缺时来得大。这种特定的超额能量可以表现为具有某种反引力效应。它所起的作用，应当恰如某种宇宙学常数。

① 相变：物质系统中物理、化学性质完全相同，且与其他部分有明显分界面的均匀部分称为相，不同相之间发生的转变称为相变。

爱因斯坦在尝试构建稳态宇宙模型时，在广义相对论中引入了宇宙学常数。然而，在这种情况下宇宙应当已处于膨胀之中。因此，宇宙学常数的斥力效应会使宇宙以不断增长的速率膨胀。即使在物质粒子多于平均值的那些区域内，有效宇宙学常数的斥力还是会超过物质的吸引力。所以，这些区域也会以某种加速暴胀的方式膨胀。

随着宇宙的膨胀，物质粒子间的距离便越来越远。结果应当是留下一个不断膨胀中的宇宙，且其中几乎不含任何粒子[①]。宇宙仍然会处于过冷态，而力与力之间的对称性并没有发生破缺。宇宙中的任何不规则性正是因为膨胀而被抹平了，这种情况就像气球表面的褶皱，一旦把气球吹胀，这些褶皱就会被抹平掉。因此，宇宙目前的平滑、均匀状态，便可以从多种不同的非均匀初始状态演化而来。膨胀的速率也会不断逼近刚好能使宇宙避免再度坍缩所需要的临界值。

不仅如此，暴胀的概念还可以用来解释宇宙中为什么会有如此多的物质。宇宙中，在我们所能观测到的区域内大约有10^{80}个粒子[②]。所有这些粒子都是从哪里来的呢？答案是，根据量子力学，粒子能以粒子/反粒子对[③]的形式由能量产生。但是，这马上又会引出能量应来自何处的问题。答案是宇宙的总能量恰好为零。

宇宙中的物质是由正能量生成的。然而，由于引力的存在，所有的物质都会彼此互相吸引。对两块相互靠得很近的物质来说，它们所具有的能量要比同样两块物质相距很远时的能量来得小。这是因为把它们分开来一定要

① 原文如此。应理解为物质密度接近零。

② 原文为“1080个粒子”，有误。

③ 粒子/反粒子：所有原子核以下层次的粒子，都有与其质量、寿命、自旋、同位旋相同，但电荷、重子数、轻子数、奇异数等量子数异号的粒子存在，称为所对应粒子的反粒子。

消耗能量。你必须抗拒引力的作用，使它们不致被吸引在一起。因此，从某种意义上说，引力场具有负能量。就整个宇宙而言，可以证明这种负引力能恰好与物质的正能量相抵消。所以，宇宙的总能量为零。

既然零的两倍还是等于零，那么要是能使宇宙中的正物质能增大一倍，又使负引力能也增大一倍，则不会违反能量守恒定律[①]。在宇宙的正常膨胀期内，随着宇宙的变大，物质的能量密度会减小，因此上述情况便不会发生。然而，在暴胀时确实会出现这种情况，原因在于尽管宇宙在膨胀，但过冷态的能量密度始终保持不变。当宇宙的尺度增大一倍时，正物质能和负引力能都增加了一倍，于是总能量仍然保持为零。在暴胀阶段，宇宙的尺度极度增大。因此，能用于生成粒子的总能量值会变得非常之大。正如古思所说的那样："常说世间不存在诸如免费午餐之类的东西。但是，宇宙却是最为丰盛的免费午餐。"

① 能量守恒定律：能量可以从一种形式转换为另一种形式，从一个物体传递给另一个物体，在这种转换和传递过程中，各种形式能量之总和保持不变。

暴胀的结局

今天，宇宙并不以暴胀方式在膨胀。所以，必然存在某种机制，以能消去非常大的有效宇宙学常数。它会改变膨胀的速率，从加速膨胀变为在引力影响下的减速膨胀，而后者正是今天所看到的情况[②]。可以预见到的情况是，随着宇宙的膨胀和冷却，力与力之间的对称性最终会出现破缺，正如过冷态水最终总是会结冰一样。那时，未破缺的对称性状态的多余能量会被释放出来，并再度使宇

② 鉴于20世纪末暗能量的发现，目前流行的观点认为宇宙正处于加速膨胀之中。

宙升温。之后，宇宙会继续膨胀并冷却，情况与热大爆炸模型完全一样。但是，宇宙为什么恰好以临界速率在膨胀？为什么宇宙的不同区域有着相同的温度？对此现在应当给出解释。

在古思的原始思想中，他假设转变为对称性破缺的过程是突然出现的，这种情况有点像在极冷的水中冰晶的显现。他的观念是，就对称性破缺后的新相而言，其中的"泡"应当是从旧相中生成的，情况就像是沸水中冒出的蒸汽泡。古思推测这些泡会膨胀，它们会互相碰在一起，直到整个宇宙进入新相。我和其他一些人曾经指出，这里的困难在于宇宙膨胀的速度是很快的，那些泡会迅速地彼此远离，而不会互相并合。宇宙最终应当处于某种高度非均匀状态，在某些区域中会保持不同力之间的对称性。这样的一种宇宙模型与我们今天所看到的情况就不相一致了。

1981年10月，我曾赴莫斯科参加一次有关量子引力的会议。会后，我在史天堡天文研究所举行了一次研讨会，内容涉及暴胀模型和它的一些问题。听众中有一位年轻的俄罗斯人安德雷•林德。他认为，如果那些泡非常之大，就可以回避有关泡无法并合的困难。如是，则可以把宇宙中我们所处的区域包含在单个泡之内。为使这一思想能行之有效，在这个泡的内部，从对称到对称破缺的变化过程必须非常缓慢地进行，不过根据大统一理论，要做到这一点是完全有可能的。

林德关于对称性缓慢破缺的思想是非常出色的，不过我曾指出他的那些泡会比当时宇宙的尺度还要大。我说明了可以换一条思路，即对称性会在所有的地方同时破缺，而不仅仅是在泡的内部发生破缺。在这种情况下便会得出如我们所观测到的那种均匀宇宙。为了解释宇宙为什么会沿着既有的方式演化，缓慢对称破缺模型是一种不错的尝试。但是，我和其他一些人已经证明，它所预言的微波背景辐射的变化要比实测结果大得多。还有，后来的一些工作也对早期宇宙中是否会存在恰当类型的相变提出了质疑。林德在1983年采用了一种更好的模型，称为混沌暴胀模型。这种模型与相变无关，而且所给出的背景辐射变化之幅度亦恰到好处。这种暴胀模型表明，宇宙目前的状态，可以由大量各不相同的初始结构演化而成。然而，并不能说每一类初始结构都应当会演化成我们所观测到的那种宇宙。所以，即使是暴胀模型也并未告诉我们，为了生成现在观测到的宇宙，为什么其初始结构就应该如此。我们必须转而用人择原理来寻求解释吗？所有这一切是否仅是一种幸运的巧合呢？那样的话似乎有点自暴自弃的味道，是对我们为理解宇宙基本秩序所寄予的全部希望的一种否定。

量子引力

为了预测宇宙应该如何起源，人们需要一些在时间起点之际能得以成立的定律。如果经典广义相对论是正确的话，那么由奇点定理可知，时间起点应当始于密度和曲率均为无穷大的一点。在这样的一点上，所有已知的科学定律全都会失效。也许可以设想，有一些新的定律在奇点处是成立的，不过在此类行为极其怪异的点上，哪怕是用公式来表述定律都非常困难，也无法通过实测来指导我们探知这些定律可能有的内容。但是，奇点定理的真实含意是，引力场变得非常强，因而量子引力效应就变得很重要：经典理论不再能很好地描述宇宙。所以，人们必须用量子引力理论来讨论宇宙的极早期阶段。下面我们将会明白，在量子理论中，一些常见的科学定律在任何场合都有可能成立，其中包括时间的起点。没有必要为奇点假设一些新的定律，因为在量子理论中无需出现任何奇点。

我们迄今还没有一种完整而又自洽的理论，以能把量子力学与引力论结合起来。但是，我们完全有把握确认这类统一理论应该具有的某些特征。其中之一便是，这种理论应该兼容费因曼根据对历史求和，并用公式来表述量子理论的思想。按照这条途径，从A点出发朝B点运动的一个粒子，并非如经典理论中所出现的那样，仅有单一的历史。现在的情况不同，粒子应该遵循时空中每一条可能的路径运动。对于每一个这样的历史，都有两个数与之相对应，一个是波的幅度，另一个则代表它在循环中的位置，即相位。

比如说，为了计算粒子通过某个特定点的概率，就要确认通过该点的每个可能的历史，并对与所有这些历史相对应的波求和，之后才能得到所需要的结果。但是，如果确实想要实现这些求和，我们便会遇到一些难以克服的技术问题。为绕开这些难题，唯一的途径是采用如下的特定处理方法：我们必须对有关粒子历史的波求和，但用以表述这些历史的并不是你我都能体验到的实时，而是虚时。

欧几里得（前330—前275）

虚时听起来也许有点像科幻小说，不过实际上它是一种有明确含意的数学概念。要想避开为实现费因曼对历史求和而在技术上出现的一些困难，我们必须采用虚时。虚时对时空有着一种奇妙的效应：时间和空间之间的区别完全不复存在。人们认为事件的时间坐标取为虚数的时空属于欧几里得时空[①]，因为度规是按正向定义的。

在欧几里得时空中，时间的方向与空间的各个方向不存在任何差别。另一方面，在实时空中，事件的时间坐标被赋以实数，因而不难发现差异之所在。时间方向处于光锥之内，而空间方向则位于光锥之外。我们可以把引入虚时视为只是一种数学手段，或者说是一种巧计，它用以就实时空来计算问题的答案。不过，也许其含意并非仅止于此。可能的情况是，欧几里得时空乃是基本概念，而我们视之为实时空者只不过是我们想象中的虚构之物。

① 欧几里得时空，以欧几里得几何为基础的四维时空，时空中的两个图形被认为是等价的，通过一系列的平移和旋转可以把一个图形变换成另一个图形。

如果我们把费因曼对历史求和的思想用于宇宙，那么现在与粒子历史相对应的就是一种代表整个宇宙历史的、完整的弯曲时空。鉴于上述技术方面的原因，必须把这些弯曲时空看作是欧几里得时空。这就是说，时间是虚的，它与空间的各个方向是不可区分的。对于一个具有某种确定性质的实时空来说，为了计算它可能出现的概率，就要在具有这种性质的虚时中，把与全部历史相对应的波相叠加。之后，才能弄清楚宇宙在实时中可能会有什么样的历史。

无边界条件

在以实时空为基础的经典引力理论中，宇宙的行为只有两种可取的方式。或者它永远存在，无始无终；或者在过去某个限定的时间，宇宙从奇点起有自己的开端。事实上由奇点定理可知，宇宙必然取第二种可能性。另一方面，在量子引力理论中还会出现第三种可能性。因为这时用的是欧几里得时空，时间方向与空间的各个方向完全处于同等地位，故时空在范围上可能是有限的，但并不存在构成边界或者边际的任何奇点。时空应当就像地球的表面，只是还多了两维。地球表面在范围上是有限的，但它并没有边界或者边缘。如果您驾船朝日落方向快速驶去，那么您不会因到达边缘而坠落，或者说不会掉入一个黑洞。我明白这一点，因为我有过环球旅行的经历。

如果欧几里得时空朝着无限远的虚时回溯，或者从某个奇点出发，那么就会出现经典理论中的同样问题，即要具体设定宇宙的初始状态。上帝也许知道宇宙是怎样诞生的，但我们不可能提出任何特定的理由，来推想宇宙会按某一种方式诞生，而不会取另一种方式。另一方面，量子引力理论则提出了一种新的可能性。在这种理论中，时空是不会有任何边界的。因此，也就无需具体设定边界处的行为。这里不会存在使科学定律失效的奇点，对时空也无边际可言，人们无需不得不求助于上帝，或者去探究某种新的定律以能为时空设定边界条件。人们可以说："宇宙的边界条件就是它没有边界。"宇宙应能做到充分自足，不会受自身之外任何事物的影响。它既不会被创造出来，也不会毁于一旦。它应当从来就是这种样子。

正是在梵蒂冈会议上，我第一次提出了这样的看法：时间和空间可能共同形成了一个面，这个面的范围是有限的，但它并没有边界或边际。然而，在我的论文中数学推演占了相当大的部分，所以当时人们并没有注意到它对宇宙创生过程中上帝所起作用的含意——对我来说也同样如此。在梵蒂冈会议期间，我还不知道如何利用无边界思想来对宇宙做出一些预言。不过，接下来的那个夏天，我是在加利福尼亚大学圣巴巴拉分校度过的。在那里，我的一位同事和朋友吉姆•哈特勒与我一起弄清楚了，如果时空无界，宇宙必须满足什么样的一些条件。

我应该强调的是，时空应该有限而无界的这种观念只是一种设想，它不可能从其他某个原理经推演而导出。就像任何别的科学理论一样，它的提出最初只是基于一些美学的或者先验的理由，但实际上的验证则在于它是否能做出一些与观测相一致的预言。然而，在量子引力框架中要确认这一点颇为不易，其原因有二。第一，我们还不能完全肯定，哪一种理论能把广义相对论和量子力学成功地结合在一起，尽管我们对此类理论必然具有的形式已取得相当多的认识。第二，任何一种模型，要能描述整个宇宙的细节情况，在数学上应当是极为复杂的，因而我们根本不可能通过计算来推知精确的预言。所以，人们不得不采取一些近似的做法——即便如此，精确预言的问题仍然相当棘手。

人们根据这种无边界的设想发现，在大多数情况下，宇宙遵循某个可能的历史而演变之机会可以忽略不计。但是，确实存在一族特定的历史，它们出现的可能性要比其他历史大得多。要是用图来表示，这些历史也许就像是地球表面，其中以北极距表示虚时；用纬圈的大小代表宇宙的空间尺度。宇宙刚诞生时位于北极，它只是一个点。随着向南运动，纬圈渐而增大，相当于宇宙随虚时在膨胀。在赤道上宇宙的尺度会达到极大；然后它会再度收缩，直至到达南极时又成为一个点。尽管在南北两极处宇宙的尺度为零，但这两个点并不是奇点，这与地球上的南北两极并无奇点之特性完全一样。在宇宙诞生之初，科学定律应当会成立，就像它们在地球南北两极成立一样。

然而，宇宙在实时中的历史看来会有很大的不同。宇宙在诞生时表现为具有某种极小的尺度，该尺度等于虚时中历史的极大尺度。然后，宇宙会在实时中膨胀，情况则与暴胀模型一样。不过，现在应当没有必要设定宇宙的生成方式，如取一种恰当类型的状态，以及通过某种途径等。宇宙会膨胀到非常大的尺度，但是最终它会再度坍缩成在实时中视之为奇点的那种模样。因此，从某种意义上说，即使我们远离黑洞，但所有的人仍然在劫难逃。只有当我们可以依据虚时来表述宇宙时，才不会出现任何奇点。

经典广义相对论的奇点定理表明，宇宙必然有一个开端，而且这个开端只能用量子理论来描述。这接下来又会引出如下的观念：在虚时中宇宙可以是有限的，但它没有边界，或者说不存在奇点。然而，一旦回到我们所生活的实时之中，奇点看来仍然是存在的。对不幸落入黑洞中的宇航员来说，他仍然会面临一种极为痛苦的结局。只有当他能够生活于虚时之中，才不会遭遇任何黑洞。

这也许会使我们想到，所谓虚时实际上就是基本时，而被我们称为实时者，只不过是我们头脑中所创造出来的某种东西。在实时中，宇宙有一个开端和一个终点，它们都是奇点，并构成时空的边界，科学定律在奇点处失效。但是，在虚时中就不存在任何奇点或边界。所以，也许被我们称之为虚时者，实际上有着更为基本的概念，而所谓实时仅仅是我们创造出来的一种概念，可用来帮助我们描述我们想象中的宇宙之模样。然而，根据第一讲中我所介绍过的思路，科学理论只是一种数学模型，它可以用来说明我们的观测结果。它仅存在于我们的脑海之中。因此，提出这样的问题是毫无意义的：哪一种是真实的，是“实”时还是“虚”时？这只不过是关于取哪一种对描述宇宙更为有用的问题。

看来，无边界设想所做出的预言是，在实时中宇宙的行为应该类似于暴胀模型。一个特别令人感兴趣的问题是，早期宇宙中对密度均匀分布的少量偏离究竟有多大。人们认为，这类偏离会导致首先形成星系，然后是恒星，最后形成像我们这样的生命。测不准原理所隐含的一个推论是，早期宇宙不可能完全均匀。相反，粒子在位置和速度上必定存在某些不确定性，或者说涨落。人们由无边界条件推知，宇宙诞生之初必然恰好具有为测不准原理所容许的最小可能的不均匀性。

因此，宇宙应当如暴胀模型所表述的那样，经历过一段快速膨胀的时期。在这段时间内，那些初始不均匀性会被放大，直至它们可以增大到足以用来解释星系的起源。所以，我们在宇宙中所观测到的一切复杂结构，都可以利用有关宇宙的无边界条件和量子力学的测不准原理来做出解释。

时空可以形成一种无边界闭合曲面的观念，同样对上帝在宇宙事务中的作用具有深远的含意。随着科学理论在描述事件时所取得的成功，大多数人渐而相信上帝容许宇宙会按照一套定律来演化。看来他不会干涉宇宙以致破坏这些定律。但是，这些定律并没有告诉我们宇宙在诞生之时看上去应该是何种模样。宇宙应当仍需仰仗上帝来上紧其发条，并选定以何种方式来启动它。只要宇宙有开端，而这个开端又是一个奇点，那么人们就可以假设宇宙乃是在某种外部力量的作用下生成的。然而，如果宇宙确实做到充分自足，不存在任何的边界或者边际，那么它就既不会被创造出来，也不会毁于一旦。宇宙应当从来就是这种样子。那么，造物主的位置又在哪里呢？

第6讲
时间的方向

为什么时间总是一直向前?

L. P.哈特利在他的《中间人》一书中写道:“过去,乃是异国他乡。那里的人行事之方式与这儿颇不相同——但是,为什么过去与未来会有如此大的差别?为什么我们所记住的只是过去,而不是未来?”换句话说,为什么时间总是一直向前?它与宇宙正在膨胀这一事实有关联吗?

C, P, T

物理学定律对过去和对未来是没有区别的。说得更确切一些,物理学定律不会因称之为C、P和T的联合动作而发生改变(C表示把粒子转变为反粒子。P的意思是取镜像,也就是左右彼此互换。T指的是把全部粒子的运动方向反过来——其效果就是做反向运动)。在所有的常规情况中,支配物质行为的物理学定律不会因动作C和P而自行改变。换言之,要是另一颗行星上的居民是我们的镜像,且由反物质组成,则那里的生命应当与我们完全一样。如果您遇到来自另一颗行星上的人,而他伸出了他的左手,那么请勿与他握手。他也许是由反物质构成的。一旦握手,你们两人会瞬间消失,并化作一片极其明亮的闪光。如果物理学定律不会因C和P的联合动作而发生改变，且不会因C、P和T的联合动作而改变，那么它们在仅有动作T时也必然不会改变的。但是，日常生活中的时间在向前与朝后两个方向之间却有着很大的差别。设想有一杯水从桌子上跌落，并在地板上摔成碎片。要是您把这一事件拍摄成一段录像，则不难区分该过程是向前还是朝后。如果您将其倒退放映，您会看到杯子的碎片突然间自行聚合且从地板上跃起，这些碎片会组合成一只完整的杯子，并跳回到桌面上。您能判定这时录像是在倒放，因为在日常生活中绝不可能观测到这类行为。要是真会出现这种事件，陶瓷厂商必会破产无疑。

时间箭头

至于我们为什么看不到破碎了的杯子会跳回到桌面上的问题，通常给出的解释是热力学第二定律不容许出现这类事件。该定律告诉我们，无序程度，或者说熵，总是随时间而增大。换句话说，这也就是墨菲定则[①]的内涵——事物总是朝坏的方向发展。桌子上一只完好的杯子处于高度有序状态，而地板上打破了的杯子是一种无序状态。因此，可以从过去桌面上一只完好无缺的杯子变为未来地板上的破碎杯子，但反其道行之则不行。

① 据说，该定则最早由美国空军上尉墨菲于1949年提出而得其名。

无序程度，或者说熵随时间而增大，是所谓时间箭头的一个例子，此类概念给时间一种方向，将过去和将来区分开来。至少存在三类不同的时间箭头。第一类，热力学时间箭头——沿着这一时间方向，无序程度或熵总是在增加的。第二类，心理学时间箭头。这个方向使我们感觉到时间在不断流逝——沿着这一时间方向，我们记住了过去，但对未来并无所知。第三类，还存在宇宙学时间箭头。沿着这一时间方向，宇宙是在膨胀，而不是在收缩。

下面我要来论证心理学箭头取决于热力学箭头，因而这两个箭头始终指向同一方向。如果对宇宙采用无边界假设，那么上述两个箭头便都与宇宙学时间箭头有关，尽管三者也许并不指向同一个方向。然而，我将要说明的是，只有当前两个箭头与宇宙学箭头相一致时才会生成智慧生命，而此类生命还能提出这样的问题：为什么无序程度随时间增加的方向与宇宙膨胀的方向相同？

热力学箭头

首先我要讨论有关热力学时间箭头的问题。热力学第二定律是基于这样一个事实：无序状态的数目要比有序状态多得多。例如，现在来考虑一盒拼图板。对于盒中的拼板来说，有一种、也仅有一种排列方式可以拼成一幅完整的图案。另一方面，要是把这些拼板作无序安放而并不构成一幅图案，则排列方式的数目是非常非常多的。

假设有一个系统，一开始处于某种有序状态，而此类有序状态为数甚少。随着时间的推移，该系统会按照物理学定律演化，它的状态也就随之发生变化。过了一段时间后，系统便会以很大的概率处于一种较为无序的状态，其原因很简单——无序状态比有序状态多得多。因此，只要系统遵循一个高度有序的初始条件，无序程度就会表现出随时间而增大的趋势。

假定在开始时拼板作有序排列，并构成一幅图案。一旦您摇动拼图盒，拼板便会呈现另一种排列方式。这很可能是一种无序排列，这时拼板并不会组成一幅应有的图案，原因很简单：无序排列的方式较之有序排列要多得多。拼板的某些组合仍然有可能构成图案的某些部分，但您摇晃拼图盒的次数越多，这类拼板组合越有可能被破坏掉。拼板终将会呈现一种完全杂乱无章的状态，这时它们不会组成任何式样的图案。因此，如果一开始这些拼板遵循处于某种高度有序状态的初始条件，那么它们的无序程度就很可能随时间而增大。

然而，现在假设上帝决定宇宙在晚期应该以某种高度有序的状态终其一生，而不管它诞生时是什么状态。那么，在早期阶段宇宙很有可能处于某种无序状态，且无序程度应当随时间而减小。您会看到破碎了的杯子自行结合在一起，并跳回到桌面上。不过，对任何正在观察那些杯子的人而言，他们会生活在无序程度随时间而减小的宇宙之中。我将要说明，这些人会有一种反向的心理学时间箭头。也就是说，他们会记着将来的事情，而对之前的事情却并无记忆。

心理学箭头

谈论人的记忆力可不是一件容易的事，因为我们并不清楚大脑的具体运作细节。不过，我们确实掌握了计算机存储器的全部工作原理。因此，下面我将讨论计算机的心理学时间箭头。我认为，可以合理地假定计算机与人的时间箭头是相同的。不然的话，人们便可以利用一台会记得明天的股票价格的计算机，在股票交易中大获收益。

就本质上说，计算机存储器是一种能对两类状态取其一的设备。金属线的超导回路可算是一个例子。如果回路中有电流通过，因为不存在任何电阻，电流便会持续不断地流动。相反，要是没有电流，回路中就会保持这种无电流状态。存储器的这两种状态可以用“1”和“0”来标识。

在某条记录录入存储器之前，存储器处于无序状态，对1和0有相同的概率。当存储器与系统互动并成为有记忆状态后，它便会根据系统的状态，明确地取上述两类状态中的一种。由此可见，存储器从无序状态转变为有序状态。但是，为了确保存储器处于正确的状态，必须消耗一定的能量。这部分能量以热的形式被消耗掉了，宇宙中的无序程度便因此而增大。可以证明，无序程度的这种增量，比存储器有序程度的增量来得大。因此，当计算机在存储器中录入一条记录时，宇宙中无序程度的总量就会增大。

计算机记住了过去，而其所取的时间方向与无序程度的增大方向是相同的。这意味着我们对时间方向的主观感觉，即心理学时间箭头，是由热力学时间箭头所确定的。这么一来，热力学第二定律便变得可有可无。无序程度随时间增大的原因在于，我们就是沿着无序程度增大的方向在量度时间。您不可能会有比这更为稳操胜算的打赌了。

宇宙的边界条件

但是，宇宙为什么就应该在时间的一个端点，即我们称之为过去的那个端点表现为高度有序的状态呢？为什么它并不是在全部时间都处于完全无序的状态呢？不管怎么说，后者的可能性看上去也许会更大一些。又为什么无序程度增大的时间方向与宇宙膨胀的方向是相同的呢？一个可能的答案是，上帝直接选定宇宙在膨胀阶段[1]之初就应该处于一种平滑而有序的状态。我们不应该试图理解这是为什么，也就是说不应去质疑上帝的理由何在，因为宇宙的开端乃是上帝之责。不过，也可以说宇宙的整个历史就是上帝的杰作。

① 膨胀阶段，宇宙学中称为膨胀相，下同。

看来，宇宙是遵循一些非常确定的规律在演化。这些规律也许受制于上帝的指令，也可能并非如此，但我们似乎有能力去发现并理解这些规律。因此，要是希望同样或者类似的一些规律在宇宙诞生之初也能成立，这难道就不合理吗？在经典广义相对论中，宇宙的开端必然是一个奇点，它在时空曲率上的密度为无穷大。在这样的条件下，所有已知的物理学定律就会全部失效。由此可见，我们不可能利用这些定律来对宇宙应会如何诞生做出预言。

宇宙在诞生之初可能就已处于一种非常平滑而有序的状态。这种状态的后随结果是必有明确的热力学时间箭头和宇宙学时间箭头，而我们所观测到的也正是这种情况。但是，同样还存在另一种可能性，即宇宙诞生时处于一种普遍呈现成团结构的无序状态。对这种情况来说，宇宙应当已经处于一种完全无序的状态，所以无序程度就不可能随时间而增大。就无序程度而言，应当存在两种可能性：或者保持不变，这时就不会有明确定义的热力学时间箭头；或者会逐渐减小，这时热力学时间箭头应当指向与宇宙学箭头相反的方向。这两种可能性中的任何一种都与我们所观测到的事实相左。

前面我已经提到，经典广义相对论预言，宇宙应该从一个奇点开始，而奇点的时空曲率为无穷大。事实上，这意味着经典广义相对论预言了它自身的垮台。一旦时空曲率变得很大，量子引力效应便会成为重要因素，这时经典理论就不再能用来对宇宙做出完好的描述。人们必须应用量子引力理论，来理解宇宙究竟是如何诞生的。

在量子引力理论中，考虑到了宇宙全部可能的历史，其中每一个历史都有两个数字与之相联系。一个代表了波的幅度，另一个则表征波的相位，即究竟是波峰还是波谷。为了给出宇宙具有某一种特定性质的概率，应确认具有这种性质的全部历史，再把表征这些历史的波叠加起来。这些历史应当是一些弯曲空间,它们代表了宇宙随时间的演化情况。我们还必须说清楚，在过去的时空边界上,宇宙可能有的那些历史应当会有怎样的行为。我们不知道,也不可能知道过去的宇宙边界条件。不过,如果宇宙的边界条件就是它没有任何边界，那么这个难题也就不存在了。换言之，所有可能的历史在范围上都是有限的,但它们没有任何边界、边际或者奇点。它们犹如地球的表面,只不过多了两维。在这种情况下,时间的起点应当是时空中一个既规则又平滑的点。这意味着,宇宙应当以一种非常平滑而有序的状态开始它的膨胀。宇宙不可能绝对均匀,否则便会违背量子理论的测不准原理。粒子的密度和速度必然都存在少量的涨落。不过,无边界条件应当隐含了这样一层意思:这类涨落会尽可能小,且与测不准原理相一致。

在诞生之初,宇宙会在一段时间内呈现指数式膨胀,或者说“暴胀”。期间,宇宙的尺度会极度增大。在这种膨胀的过程中,开始时会维持少量的密度涨落不变,但随之这种涨落就开始发展。一些区域的密度会略大于平均密度,而在多余质量的引力作用下,这些区域的膨胀会逐渐减慢。最终,此类区域的膨胀应当会停下来,并经坍缩而形成星系、恒星,以及像我们这样的生命。

宇宙最初应当处于一种平滑而有序的状态，随着时间的推移,它会变成呈现成团结构的无序状态。这可以用来解释热力学时间箭头的存在。宇宙在开始时应当呈现一种高度有序的状态,并随时间而变得越来越无序。正如我在前文中证明的，心理学时间箭头所指的方向与热力学箭头是同向的。因此，我们对时间的主观感受应当是沿着这一方向宇宙正在不断膨胀，而不是取相反的方向，也就是宇宙不会处于不断收缩之中。

时间箭头会反转吗？

但是，如果宇宙一旦停止膨胀并再度开始收缩，那又会发生些什么呢？热力学箭头的方向是否应当反过来，且无序程度会随时间而开始减小呢？对于历经膨胀阶段到收缩阶段后还活着的人来说，这会引出各种各样科幻小说式的可能性。他们是否应当看到打碎了的杯子在从地板上跃起之际会自行结合，并完整无缺地跳回到桌面上呢？他们是否应当有能力记住明天的股票价格，并在股市中发一笔横财呢？

为宇宙再度坍缩之时会发生些什么情况而担忧，这看来也许有点学究气，因为至少在接下来的100亿年内宇宙是不会开始收缩的。不过，这里有一条比较快捷的途径可以用来揭示那时将会发生些什么：跃入一个黑洞的内部。恒星坍缩而形成黑洞，这一过程与整个宇宙坍缩的晚期阶段颇为相似。因此，如果在收缩阶段中宇宙的无序程度应该减小，那么我们还可以预期到，在黑洞内部无序程度也是会减小的。所以，对于掉入黑洞的宇航员来说，也许他会在下赌注之前记住了球将滚向何处，从而能在轮盘赌中赢钱。不过，遗憾的是，他不会有多长的时间来玩此类赌博，因为黑洞内极强的引力场很快就会把他变成意大利面条了。鉴于他已陷入黑洞事件视界的幕布之后，他既无法让我们知道有关热力学箭头反转的点滴情况，甚至也不可能把所赢得的钱存入银行。

最初，我相信当宇宙再度坍缩时，无序程度应当会减小。这是因为我曾推想一旦宇宙又变得很小，它必然要回复到一种平滑而有序的状态。这就意味着，收缩阶段就像是膨胀阶段的时间反演。对处于收缩阶段中的人而言，他们的生活经历与今日之生活相比应当是颠倒的。他们是去世在前，出生在后，还会随着宇宙的收缩而变得越来越年轻。这种概念很吸引人，因为它意味着宇宙的膨胀和收缩阶段之间有一个美妙的对称。但是，人们不可能只是用这一个概念，而无视有关宇宙的其他一些概念。问题在于，这种概念究竟是无边界条件所隐含的结果，还是与那种条件并不相符呢？

前文我已提及，最初我曾以为无边界条件确实隐含了无序程度在收缩阶段应当会减小。这种观念的基础乃是就某种简单宇宙模型所做的工作，而在模型中收缩阶段看上去就像是膨胀阶段的时间反演。然而，我的一位同事唐•佩奇指出，无边界条件并不要求收缩阶段必然是膨胀阶段的时间反演。还有，我的一名学生雷蒙德•拉夫莱姆发现，只要模型稍微变得复杂一些，宇宙的坍缩过程便与膨胀大相径庭了。我意识到我犯了一个错误。事实上，无边界条件隐含了在收缩期间无序程度应当继续维持增大。无论是宇宙开始再度收缩，还是在黑洞的内部，热力学和心理学的时间箭头都不会反转。

如果您发现自己犯了一个诸如此类的错误，您又应该作何种处理呢？有些人，如爱丁顿，从来不承认他们会做错事。他们会继续去寻找一些新的、而且往往是自相矛盾的证据，来支持他们对问题的立场。另一些人的做法是，绝不无原则地支持最初的不正确观点，或者，要是他们做过什么工作的话，也只是为了证实那是有问题的。我可以列举大量这方面的事例，但我不会那样做，因为这样做的话我就太不得人心了。在我来看，如能以书面方式承认自己犯了错，这种做法就会好得多，也主动得多。爱因斯坦对此做出了很好的榜样，他就说过，在尝试构筑静态宇宙模型时所引入的宇宙学常数，是他一生中所犯的最大错误。

第7讲
万物之理

若想一劳永逸地为世间万物构筑一种完整的统一理论，那会是非常困难的。所以，我们走的是另一条路，那就是通过发现局部性理论以求得进展。这些理论对有限范围内的事物作了描述，而对其他因素的影响则不予考虑，或者以一些确定的数字作为它们的近似。例如，在化学中，我们可以计算原子间的相互作用，而无需知晓原子核的内部结构。但是，归根结底人们总是希望能找到一种完美而又自洽的统一理论，而且它应能包容作为其近似表述的所有那些局部性理论。对此类理论的探求被称为"物理学的统一性"。

爱因斯坦把他晚年的大部分时光用于探索一种统一理论，但未取得成功，不过在那个年代时机尚未成熟，因为当时人们对核力所知甚少。另外，爱因斯坦拒不相信量子力学的真实性，尽管他对量子力学的发展曾发挥了重要作用。但是，测不准原理看来正是我们生活之宇宙的一个基本特征。因此，一种成功的统一理论必须明确地能包容这一原理。

到了今天，能找到此类理论的前景似乎要好得多了，因为我们对宇宙的认识已取得长足的进步。不过，我们必须谨防过分自信，因为之前已经有过似是而非的教训。例如，在20世纪初，人们曾以为世间万物都可以用连续物质的一些性质，如弹性和热传导性来加以解释。鉴于原子结构的发现和测不准原理，这种想法便告寿终正寝。

后来又有过一次：1928年马克斯·波恩曾对访问哥廷根大学的一批人说，“就我们所知，物理学将在六个月内终结”。他的这种自信，是基于不久前狄拉克发现了支配电子的方程[①]。有人认为应当有一个类似的方程支配质子（质子是当时所知唯一的一种另类粒子），而理论物理也就到此为止了。然而，中子和核力的发现又一次给持有此类观点的人当头一棒。

① 即由英国物理学家P. A. M. 狄拉克于1928年提出的电子相对论性运动方程，从而奠定了相对论性量子力学的基础。

尽管说了上面这些话，我仍然以一种谨慎乐观的态度相信，现在有理由说我们可能已经接近探知自然界终极规律的目标。今天，我们已掌握了若干局部性理论。我们有了广义相对论，这是有关引力的局部性理论，以及还有支配弱力、强力和电磁力的局部性理论。后三种理论可以合并为所谓的大统一理论。不过这类理论并不非常令人满意，因为它们没有把引力包括在内。要想找到一种理论，以能把引力与其他几种力统一起来，其主要困难在于广义相对论乃是一种经典理论。这就是说，它并不包容量子力学的测不准原理。相反，其他三种局部性理论都与量子力学紧密地联系在一起。因此，必须做的第一步工作是，要把广义相对论与测不准原理结合起来。正如我们已看到的那样，这样做可以导出一些很值得注意的结论，如黑洞并非黑不可知，宇宙是完全自足的，而且没有边界。

麻烦在于，测不准原理意味着即使是完全真空的空间也充满了虚的粒子和反粒子对。这些虚粒子对会具有无限大的能量，而这意味着它们的引力作用会使宇宙高度弯曲成无限小的尺度。

在其他一些量子理论中，也会出现一些颇为类似的、看上去很荒唐的无限大问题。不过，在这些理论中，此类无限大可以通过一种所谓重正化[①]的处理方法而不复出现。这种方法涉及对理论中粒子的质量和力的强度进行调整，而调整的范围也是无限大。尽管这种方法就数学上来说颇为令人生疑，但在实际应用上看来却是有效的。人们已经利用这种方法做出了一些预言，而且以异乎寻常的精确度与观测结果相一致。但是，从力图找到一种完美理论的观点来看，重正化有一个严重的缺陷。一旦从无限大中扣除无限大，那么您想要什么答案就可以取得什么答案。这意味着，从这种理论不可能预言质量和力的强度之实际数值。相反，它们不得不通过与观测结果间的拟合来加以选定。对广义相对论来说，只有两个量是可以调整的，那就是引力强度和宇宙学常数。然而，调整这两个量尚不足以避开所有的无限大。为此，有人提出了一种理论，这种理论看来能对某些量（如时空曲率）做出预言，这些量尽管实际上为无穷大，但它们是可以观测的，且测定值必然是有限的。为了解决这一难题，1976年有人提出了一种称之为“超引力”的理论。这种理论本质上就是广义相对论，只是补充了一些粒子。

① 重正化：克服量子理论中的发散困难，使理论计算得以顺利进行的一种理论处理方法。

在广义相对论中，引力可以看作起因于一种自旋为2

的粒子，这就是引力子。上述理论的思想是，应增加自旋为3/2、1/2和0的其他几种与之不同的新粒子。于是，从某种意义上说，所有这些粒子都可认为是同一种“超粒子”的不同表现。对于自旋为1/2和3/2的虚粒子/反粒子来说，它们应当具有负能量，而这种负能量往往会与自旋为0、1和2的虚粒子对的正能量相抵消。这样一来，许多可能的无限大也就不复存在，但令人担心的是有些无限大仍然可能留存下来。不过，要想确认是否仍留下某些无限大而未被消除，所需要的计算工作量非常大，且难度很高，因而没有人打算进行此类计算。即使用计算机来算，估计至少也得花上四年时间。在计算中至少出错一次，或者也许多次出错，这种可能性是很大的。所以，若想知道某人得出的答案是正确的，那么必得另有人重复这项计算，并能取得相同的结果，而要做到这一点看来是不太可能的。

鉴于上述难题的存在，另一种与之不同的见解便应运而生，那就是弦论。在此类理论中，基本客体不再是在空间中只占了一个点的粒子，而代之以某种有长度却无其他维度的东西，犹如一条无限细的弦圈。在每一瞬刻，一个粒子仅占了空间中的一点。所以，它的历史便可以用时空中的一条线来表示，称为“世界线”。另一方面，在每一瞬刻，一条弦则占了空间中的一条线，因而它在时空中的历史是一个二维曲面，称为“世界面”。此类世界面上的任意一点，可以用两个数来描述，一个表征时间，另一个则表征点在弦上的位置。弦的世界面乃是一个圆柱面，或者说一根管子。管子的截面是一个圆，它代表了某一特定时刻弦所处的位置。

两段弦可以连接起来，并合成一条弦，这有点像两条裤腿连接成一条裤子。类似地，一条弦也可以分割成两条弦。在弦论中，先前被视为粒子的客体，现在可想象为沿着弦传播的波，就像是冲水管上的波。一个粒子因另一个粒子的作用而被发射或者被吸收，与之相应的就是弦的分割或者连接。例如，与太阳对地球的引力作用相对应的便是H形的水管或导管。弦论多少有点像自来水管道。在H形结构两个竖直管道上的波，对应于太阳和地球上的粒子，而与水平连接管上的波所对应的便是在两者之间传播的引力。

弦论有着一段异乎寻常的历史。最早，它是在20世纪60年代末被虚构出来的，当时有人试图找到一种理论来描述强力。它的构想是，像质子和中子这样的粒子，可以看作为一条弦上的波。粒子间的强力对应于一些弦段，它们游走于其他一些弦段之间，这种情况有点像蜘蛛网。为了用这种理论来说明粒子间强力的观测值，那些弦必须像橡胶带一样，并能承受10吨左右的拉力。

1974年，约埃尔·歇克和约翰·施瓦茨发表了一篇文章，在文中他们证明了弦论可以用来描述引力，条件是弦的张力必须要大得非常多——约达10^{39}吨。在常见的长度范围内，弦论所作的一些预言与广义相对论的预言完全一致，然而在非常小的距离（小于10^{-33}厘米）上两者就迥异了。不过，他们的工作并没有引起太大的注意，因为正是在那一时间前后大多数人抛弃了原始的、有关强力的弦论。歇克去世时景况甚为凄惨。他得的是糖尿病，在处于昏迷状态之际周围没有人为他注射胰岛素。这么一来，孤零零的施瓦茨几乎成了弦论的唯一支持者；然而，今天提出的弦张力之数值还要大得多。

到了1984年，人们突然间再度对弦产生了兴趣，这大概有两方面的原因。其一，人们在证明超引力是有限的，或者说要证明它可以用来解释我们已观测到的那几类粒子方面，实际上并没有取得多大的进展。其二，约翰•施瓦茨和迈克•格林在他们所发表的一篇论文中证明，弦论也许能用来解释内禀左手征粒子的存在，而我们所观测到的某些粒子便具有这种特征。不管是什么原因，许多人很快开始从事弦论方面的工作。一种称为异型弦的新版本弦理论发展起来了，而且看来很有可能用它来解释我们所观测到的各类粒子。

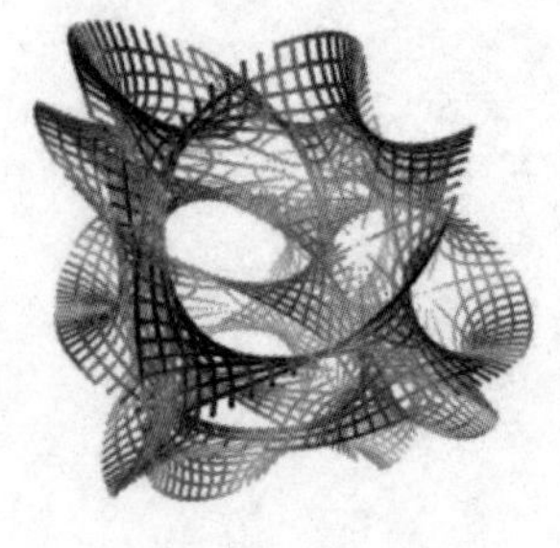

空间维度

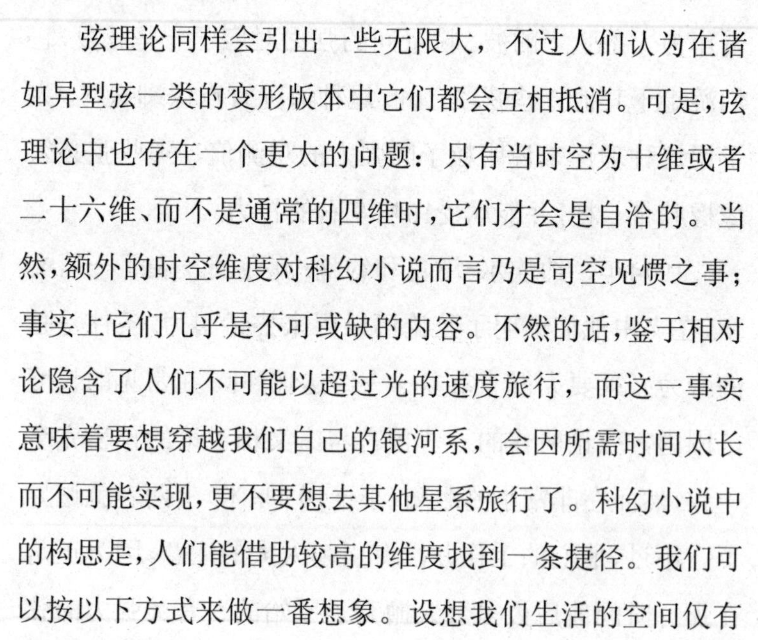

弦理论同样会引出一些无限大，不过人们认为在诸如异型弦一类的变形版本中它们都会互相抵消。可是，弦理论中也存在一个更大的问题：只有当时空为十维或者二十六维、而不是通常的四维时，它们才会是自洽的。当然，额外的时空维度对科幻小说而言乃是司空见惯之事；事实上它们几乎是不可或缺的内容。不然的话，鉴于相对论隐含了人们不可能以超过光的速度旅行，而这一事实意味着要想穿越我们自己的银河系，会因所需时间太长而不可能实现，更不要想去其他星系旅行了。科幻小说中的构思是，人们能借助较高的维度找到一条捷径。我们可以按以下方式来做一番想象。设想我们生活的空间仅有二维，而且它是弯曲形的，就像是甜甜圈或说一个环形曲面的表面。如果您位于环的某一边上，并打算去另一边上的某一点，那么您必得绕着环走。然而，要是您能做到在第三维中旅行，那么您就可以抄近路笔直横穿过去。

如果所有这些额外的维度确实存在，那我们为什么对之毫无察觉呢？为什么我们所看到的只是三维空间和一维时间呢？对此的解释是，其他的维度均弯曲在范围极小的空间内，其尺度大约只有100万亿亿亿分之一英寸。这样的尺度实在是太小了，以至于无法被我们所觉察。我们所看到的只是三维空间和一维时间，而且此类时空是完全平直的。这种情况可比作为橙子的表面：如果您在近距离观察，整个表面是弯曲的，而且布满皱纹；然而要是从远处看，您就看不到那些隆起的结构，它看上去显得很平滑。对于时空来说情况也是如此。在非常小的尺度上它是十维的，而且高度弯曲。但是在较大尺度上，您看不到弯曲，或者说看不到额外的维度。

如果这幅图像是正确的话，对于想要成为空间旅行者的人来说那可是坏消息。额外的维度实在是太小了，空间飞船无法得以通过。然而，这又引起了另一个重要问题。为什么某些维度应该卷成一个小球，而不是所有的维度都是如此呢？据推测，极早期宇宙中所有维度都应当是高度弯曲的。那么为什么三维空间和一维时间展平了，而其他维度仍维持紧卷状态呢？

一个可能的答案是人择原理。为了发展出像我们这样的复杂生命，两维空间看来是不够的。例如，对于生活在一维地球上的两维人来说，两个人若要想彼此穿越而过,那么一个人就必须从另一个人的身上攀爬过去。如果一个两维生命吃了什么东西,那它是不可能完全消化的,它不得不把残留物吐出来，而且吐出的通道与吞食通道是一样的,这是因为要是有一条通道贯穿整个生命体,那么它会把这个生命分割成两个不相连的部分，而我们的两维生命也就解体了。类似地,想要理解两维生命体内如何才能进行某种血液循环亦很困难。当空间的维度大于三时,同样会出现一些问题。这时两个物体之间的引力随距离的增大而减小,会比三维空间中减小得更快。由此引起的一个严重后果是,行星(如地球)绕太阳的运动轨道不会处于稳定态。因诸如其他行星的引力所造成的极微小的扰动,会使地球沿螺旋形轨道远离或跌入太阳。这么一来我们要么会彻底冻结,要么就被焚烧殆尽。实际上,这样一种引力随距离而变的特性，同样会使太阳处于不稳定态。太阳或者会分崩离析，或者会经坍缩而形成黑洞。无论哪种情况,作为地球上生物之光和热的源泉,太阳就没有太大的用处了。在较小尺度上,使原子中电子绕核作轨道运动的电力，也会表现出与引力相同的变化特性。于是,所有的电子都会从原子中逃逸出去,或者按螺旋式运动掉入原子核。不管出现何种情况,原子都不可能保持现在我们所知道的那种模样。

看来,生命——至少是我们所知道的那种生命,显然只能存在于特定的时空区域之中，它们有三维空间和一维时间,而且都不会因卷曲而变得非常小。这意味着人们可以援引人择原理，不过前提条件是可以证明弦理论至少确能容许在宇宙中存在此类区域。而且，看来每一种弦理论确实都容许此类区域的存在。宇宙中很可能存在其他的区域，或者还存在别的宇宙（且不论这意味着什么），而在这些宇宙中所有的维度都卷曲得很小，或者其中有四个以上的维度是近乎平直的。不过，在此类区域中不会有任何智慧生物存在，而对有效维度之不同个数的观测也就无从谈起了。

除了时空所表现出的维度之个数这个问题外，弦论还存在其他一些难点，它们必须先行解决，之后才能在欢呼声中成为物理学的终极统一理论。我们尚不清楚所有的无限大是否会互相抵消，或者还不知道如何严格地把弦上的波与我们观测到的粒子的具体类型联系起来。尽管如此，在接下来的几年内找到这些问题答案的可能性还是存在的，而到20世纪末我们将会知道弦论是否确实是人们探求已久的物理学之统一理论[①]。

① 事实上，迄今尚未就此得出令人信服的明确结论。

实际上能否存在一种适用于宇宙万物的统一理论呢？还是我们只是在追逐某种海市蜃楼式的幻象呢？这里看来存在三种可能性：

1. 确实存在一种完美的统一理论，如果我们有足够的智慧，那总有一天会找到它。

2. 不存在适用于宇宙的任何终极理论，有的只是一系列理论，它们对宇宙的描述越来越精确，但这一过程永无止境。

3. 不存在适用于宇宙的任何理论。超出一定范围后事件是不可预知的，它们只以一种随机而又任意的方式出现。

一些人赞成第三种可能性，理由是如果存在一整套完美的定律，那就会侵犯上帝改变主意和干预世界的自由。这有点像一则古老的悖论：上帝是否有能力创造出一块连他自己也无法举起的大石头呢？但是，上帝也许想要改变主意的构想，乃是圣奥古斯丁所指出的一例谬误：把上帝想象为是一种适时存在的生物。时间只是上帝创造的宇宙中的一种性质而已。据此推测，上帝在创造宇宙之时已经知道他的意图何在。

随着量子力学的面世，我们渐而意识到对事件的预言不可能做到绝对准确，而总是存在某种程度的不确定性。要是有人愿意，那他可以把这种随机特性归因于上帝的干预。但是，这应当是一种非常奇特的干预。没有任何证据表明，这种干预有着明确的目的。事实上，如果有的话，它就不会是随机的。在当今年代，鉴于重新明确了科学目标，我们实际上已经排除了第三种可能性。我们的目的在于建立起一套用公式表示的定律，从而使我们能在测不准原理所限定的精度内对事件做出预言。

第二种可能性涉及到在无限过程中越来越精确的一系列理论，这种情况与迄今我们的全部经历是相一致的。在许多场合中，我们提高测量工作的灵敏度，或者开展新一类的观测，目的只是在于揭示新的现象，而此类现象是现有理论无法预言的。有鉴于此，我们必须发展出更为先进的理论。所以，对现有的那些大统一理论来说，如果它们经更大、更强的粒子加速器检验而不再成立，那也用不着为此而过于大惊小怪。确实，如果我们的预期目标是它们不会失效，那么花那么多钱去建造更强大的设备就不会有太大的意义了。

然而，引力好像有可能为这种“盒中套盒”式的序列提供某种限制。如果有一个粒子，它的能量超过所谓普朗克能量，即10^{19}GeV①，那么它的质量便会高度密集，结果是切断了它与宇宙其余部分的联系，形成一个微小的黑洞。据此，随着所涉及的能量越来越高，这种不断精确的理论序列好像确实应该有一个极限。对宇宙来说，应该存在某种终极理论。当然，目前我们在实验室内所能产生的能量充其量也就在1GeV左右，这与普朗克能量相距甚远。为了跨越这么大的鸿沟，应当需要一台比太阳系还要大的粒子加速器。以现有的经济实力来说，想获得建造这样一台加速器所需要的经费是完全不可能的。

① 原文为“1019GeV”，有误。

不过，宇宙极早期阶段乃是此类能量必然曾经出现过的竞技场。以我所见，现在是一个很好的时机，那就是对早期宇宙的研究，以及对数学自洽性的要求，将会引导我们在20世纪末得出一种完美的统一理论——前提始终是我们不会先行自我毁灭。

如果我们真的发现了一种有关宇宙的终极理论，那又应当意味着什么呢？在我们为认识宇宙而奋斗的历史进程中，漫长而又辉煌的一章会因取得这一成果而划上句号。不过，这也会使公众对那些支配宇宙的定律之理解发生革命性的转变。在牛顿时代，受过教育的人有可能掌握人类知识之全貌，至少也能知其概况。但是，之后随着科学的不断发展，要想做到这一点已不再可能了。理论总是为了说明新的观测结果而处于不断变化之中。这种理论绝不可能通过适当的消化或者简化，以使普通大众得以理解它们。您必须成为一名专家，而即便如此，您也只能有望做到对一小部分科学理论在一定程度上有所领会。

还有，科学进步的速度非常之快，人们在中学或者大学里学到的那些知识始终是稍嫌过时；只有为数不多的人才能跟得上快速进展的知识前沿。而且，他们不得不为之投入自己的全部时间，并在某个不大的领域内从事专业性研究。至于其余的绝大多数人，对于不断取得的进展以及由此引发的激情则知之甚少。

要是七十年前的爱丁顿所言属实，则那时仅有两个人能理解广义相对论。时至今日，数以万计的大学毕业生都能理解这一理论，还有几百万人至少熟悉其思想。一旦发现了完美的统一理论，那么要想按同样方式对它进行消化或者简化，也只是一个时间问题。届时，中学里就可以讲授这种理论，至少可知其梗概。那时，我们就都能够对支配宇宙，并对我们的存在起决定作用的那些定律有所了解。

爱因斯坦曾经提出过这样一个问题："在构建宇宙之时，上帝有多大的选择余地？"如果无边界设想是正确的话，上帝在选定初始条件时就没有任何自由了。当然，他应当仍有选择支配宇宙之定律的自主权。不过，这里也不可能真有多大的自由选择的余地。很可能只存在一种或者少数几种完美的统一理论，它们是自洽的，而且容许有智慧生命存在。

即使只存在一种可能的统一理论，且这种理论又只表现为一套规律和相应的方程组合，我们还是可以就上帝的本性发问。是什么因素能把灵感注入这些方程，并创造出能用这些方程来加以描述的宇宙呢？通常用于构筑某种数学模型的科学途径，并不能回答这样一个问题：为什么应该存在一种能用该模型来描述的宇宙？为什么宇宙会克服种种麻烦以求存在呢？统一理论是否如此使人信服，以至于它自身的实现不可避免？或者它的确需要一位造物主，而如果正是如此，那么除了为宇宙的存在负责之外，上帝是否还会对宇宙施以什么影响？还有，又是谁创造了上帝呢？

迄今为止，大多数科学家都一直太过专注于发展一些新的理论，以求描述宇宙的真貌，而没有去探问为什么的问题。另一方面，对于那些探求宇宙为何如此的人（即哲学家）来说，他们的工作无法跟上科学理论的进展。在18世纪，哲学家们把包括科学在内的人类全部知识，当作他们的研究范围。他们曾经讨论了诸如“宇宙有开端吗”之类的问题。然而，到了19和20世纪，除少数科学家外，对于哲学家和其他任何人来说，科学在学术内容和数学方法上变得过于深奥。哲学家们大大缩小了他们的探索范围，而正因为如此，20世纪最著名的哲学家维特根斯坦曾说道：“对哲学而言，唯一还可以做的事就只剩下分析语

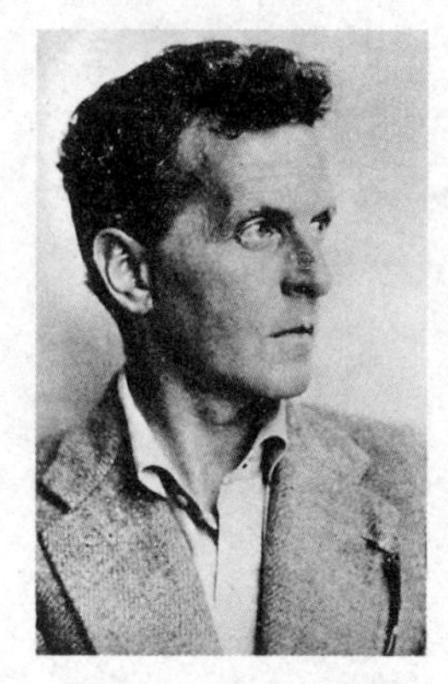
维特根斯坦 (1889—1951)

言了。”从亚里士多德到康德，哲学的伟大传统之没落竟到了如此凄惨的境地！

不过，一旦我们确实发现了一种完美的理论，那么应该及时让每个人理解其主要原理，而不只限于少数科学家。那时，我们所有的人都能参与对宇宙为什么存在这一问题的讨论。如果我们找到了这一问题的答案，那将会是人类理性的终极胜利。因为到那个时候，我们就会知道上帝的意向了。

名词对照表

absolute zero 绝对温标零度

anthropic principle 人择原理

antigravity 反引力

antimatter 反物质

Aristotle 亚里士多德

arrows of time 时间箭头

Atomic bomb project 原子弹计划

background radiation 背景辐射

balloon model of expanding universe 膨胀宇宙的气球模型

Bardeen, Jim 巴丁，吉姆

beginning of time 时间起点

Bell Labs 贝尔实验室

Bell, Jocelyn 贝尔，乔丝琳

Bentley, Richard 本特利，理查德

Bekenstein, Jacob 贝肯斯坦，雅科布

big bang theory 大爆炸理论

black hole 黑洞

blue-shift 蓝移

Bondi, Hermann 邦迪，赫尔曼

Born, Max M. 波恩，马克斯•M.

boundary condition 边界条件

brightness of stars 恒星亮度

Carter, Brandon 卡特，布兰登

Catholic church 天主教会

Cerencov radiation 切伦科夫辐射

Chandrasekhar limit 昌德拉塞卡极限

Chandrasekhar, Subrahmanyan 昌德拉塞卡，苏布拉马尼扬

chemical element 化学元素

Christian church 基督教会

City of God《上帝之城》

classical theory 经典理论

cold star 无能源恒星

Copernicus, Nicholas 哥白尼，尼古拉

cosmic censorship hypothesis 宇宙监督假设

cosmological constant 宇宙学常数

cosmology 宇宙学

creation of the universe 宇宙创生

Cygnus X-1 天鹅X-1

density 密度

Dicke, Bob 迪克，鲍勃

direction of time 时间方向

distance of stars 恒星距离

Doppler effect 多普勒效应

Earth 地球

eclipse 食，交食

Eddington, Sir Arthur 爱丁顿，亚瑟爵士

Einstein 爱因斯坦

entropy 熵

event horizon 事件视界

expanding universe 膨胀的宇宙

Feynman's idea 费因曼思想

fixed star 恒星

flood 洪水

formation of stars 恒星形成

Friedmann, Alexander 弗里德曼，亚历山大

Friedmann model 弗里德曼模型

galactic gas cloud 星系气体云

galaxy 星系

Galilei，Galileo 伽利略，加利莱奥

gamma ray 伽马射线

Gamow George 伽莫夫，乔治

general relativity 广义相对论

Genesis 创世记

Gold, Thomas 戈尔德，托马斯

grand unified theory 大统一理论

gravitational wave 引力波

gravity 引力

Green, Mike 格林，迈克

Guth, Alan 古思，艾伦

Hartle, Jim 哈特勒，吉姆

Hartley, L. P. 哈特利，L. P.

Helium 氦

Hewish, Anthony 休依什，安东尼

hot big bang model 热大爆炸模型

Hoyle, Fred 霍伊尔，弗雷德

Hubble, Edwin 哈勃，埃德温

Hydrogen 氢

hydrogen bomb 氢弹

Ice Age 冰河期

imaginary time 虚时

infinite time 无限时间
infinite universe 无限宇宙
infinity 无穷
inflationary model 暴胀模型
Israel, Werner 伊斯雷尔，沃纳

Jupiter 木星

Kant 康德
Kerr, Roy 克尔，罗伊
Khalatnikov, Isaac 哈拉特尼柯夫，伊萨克
Laplace, Marquis de 拉普拉斯侯爵
law of universal gravitation 万有引力定律
Lifshitz, Evgenii 利夫希茨，叶夫根尼
light cone 光锥
light spectrum 光谱
Linde, Andrei 林德，安德雷
luminosity of stars 恒星光度

magnetic force 磁力
Mars 火星
matter 物质
Mercury 水星
Michell, John 米歇尔，约翰
microwave detector 微波探测器
Milky Way galaxy 银河系
MIT 麻省理工学院
motion 运动
Mount Palomar Observatory 帕洛马山天文台
Murphy's Law 墨菲定则

neutron star 中子星
Newton 牛顿
no boundary condition 无边界条件
Nobel Prize 诺贝尔奖

observable universe 可观测宇宙
Olbers, Heinrich 奥伯斯，海因里希
Oppenheimer, Robert 奥本海默，罗伯特

particle theory of light 光的粒子说
Pauli exclusion principle 泡利不相容原理
Peebles, Jim 皮伯尔斯，吉姆
Penrose, Roger 彭罗斯，罗杰
Penzias, Arno 彭齐亚斯，阿尔诺
phase transition 相变
Philosophical Transactions of the Royal Society of London《伦敦皇家学会哲学学报》
physics 物理学
Planck energy 普朗克能量

Planck's quantum principle 普朗克量子原理

planetary orbit 行星轨道

Pluto 冥王星

Pole star 北极星

Porter, Neil 波特，尼尔

positive matter energy 正物质能量

primordial black hole 原初黑洞

Principia Mathematica Naturalis Causae 自然哲学的数学原理

prism 棱镜

psychological arrow 心理学箭头

Ptolemy 托勒密

pulsar 脉冲星

quantum gravitational effect 量子引力效应

quantum gravity 量子引力

quantum mechanical uncertainty principle 量子力学测不准原理

quantum mechanics 量子力学

quantum theory 量子理论

quasar 类星体

radar 雷达

radial wave 辐射波

real time 实时

red-shift 红移

renormalization 重正化

Robertson, Howard 罗伯逊，霍华德

Robinson, David 鲁宾逊，戴维

Ryle, Martin 赖尔，马丁

Saturn 土星

Scherk, Joël 歇克，约埃尔

Schmidt, Maarten 施密特，马尔滕

Schwarz, John 施瓦茨，约翰

Schwarzchild, Karl 史瓦西，卡尔

Schwarzchild solution 史瓦西解

second law of thermodynamics 热力学第二定律

selection principle 选择原理

singularity 奇点

space-time continuum 时空连续

space-time dimensions 时空维度

St.Augustine 圣奥古斯丁

Stadia 斯达地

Starobinsky, Alexander 斯塔罗宾斯基，亚历山大

star 恒星

steady state theory 稳恒态理论

Sternberg Astronomical Institute 史天堡天

文研究所

string theory 弦(理)论

strong force 强力

supercool water 过冷态水

symmetry of water 水的对称性

System of the World 宇宙体系论

Taylor, John 泰勒,约翰

theory of relativity 相对论

thermodynamic arrow 热力学箭头

Thorne, Kip 托恩,基普

uncertainty principle 测不准原理

unified force 统一力

unified theory 统一理论

Vatican 梵蒂冈

Venus 金星

Walker, Arthur 沃克,亚瑟

wave theory of light 光的波动说

weak force 弱力

Weekes, Trevor 威克斯,特雷弗

Wheeler, John 惠勒,约翰

white dwarf 白矮星

Wilson, Robert 威尔逊,罗伯特

Wittgenstein 维特根斯坦

world-line 世界线

world-sheet 世界面

worm hole 虫洞

X ray emission X射线发射

Zeldovich, Yakov 泽利多维奇,雅科夫

STEPHEN W. HAWKING

SPECIAL ANNIVERSARY EDITION

THE THEORY OF EVERYTHING

THE ORIGIN AND FATE OF THE UNIVERSE

CONTENTS

INTRODUCTION

In this series of lectures I shall try to give an outline of what we think is the history of the universe from the big bang to black holes. In the first lecture I shall briefly review past ideas about the universe and how we got to our present picture. One might call this the history of the history of the universe.

In the second lecture I shall describe how both Newton's and Einstein's theories of gravity led to the conclusion that the universe could not be static; it had to be either expanding or contracting. This, in turn, implied that there must have been a time between ten and twenty billion years ago when the density of the universe was infinite. This is called the big bang. It would have been the beginning of the universe.

In the third lecture I shall talk about black holes. These are formed when a massive star or an even larger body collapses in on itself under its own gravitational pull. According to Einstein's general theory of relativity, anyone foolish enough to fall into a black hole will be lost forever. They will not be able to come out of the black hole again. Instead, history, as far as they are concerned, will come to a sticky end at a singularity. However, general relativity is a classical theory—that is, it doesn't take into account the uncertainty principle of quantum mechanics.

In the fourth lecture I shall describe how quantum mechanics allows energy to leak out of black holes. Black holes aren't as black as they are painted.

In the fifth lecture I shall apply quantum mechanical ideas to the big bang and the origin of the universe. This leads to the idea that space-time may be finite in extent but without boundary or edge. It would be like the surface of the Earth but with two more dimensions.

In the sixth lecture I shall show how this new boundary proposal could explain why the past is so different from the future, even though the laws of physics are time symmetric.

Finally, in the seventh lecture I shall describe how we are trying to find a unified theory that will include quantum mechanics, gravity, and all the other interactions of physics. If we achieve this, we shall really understand the universe and our position in it.

LECTURE *1*

IDEAS ABOUT THE UNIVERSE

As long ago as 340 B.C. Aristotle, in his book *On the Heavens*, was able to put forward two good arguments for believing that the Earth was a round ball rather than a flat plate. First, he realized that eclipses of the moon were caused by the Earth coming between the sun and the moon. The Earth's shadow on the moon was always round, which would be true only if the Earth was spherical. If the Earth had been a flat disk, the shadow would have been elongated and elliptical, unless the eclipse always occurred at a time when the sun was directly above the center of the disk.

Second, the Greeks knew from their travels that the Pole Star appeared lower in the sky when viewed in the south than it did in more northerly regions. From the difference in the apparent position of the Pole Star in Egypt and Greece, Aristotle even quoted an estimate that the distance around the Earth was four hundred thousand stadia. It is not known exactly what length a stadium was, but it may have been about two hundred yards. This would make Aristotle's estimate about twice the currently accepted figure.

The Greeks even had a third argument that the Earth must be round, for why else does one first see the sails of a ship coming over the horizon and only later see the hull? Aristotle thought that the Earth was stationary and that the sun, the moon, the planets, and the stars moved in circular orbits about the Earth. He believed this because he felt, for mystical reasons, that the Earth was the center of the universe and that circular motion was the most perfect.

This idea was elaborated by Ptolemy in the first century A.D. into a complete cosmological model. The Earth stood at the center, surrounded by eight spheres, which carried the moon, the sun, the stars, and the five planets known at the time: Mercury, Venus, Mars, Jupiter, and Saturn. The planets themselves moved on smaller circles attached to their respective spheres in order to account for their rather complicated observed paths in the sky. The outermost sphere carried the so-called fixed stars, which always stay in the same positions relative to each other but which rotate together across the sky. What lay beyond the last sphere was never made very clear, but it certainly was not part of mankind's observable universe.

Ptolemy's model provided a reasonably accurate system for predicting the positions of heavenly bodies in the sky. But in order to predict these positions correctly, Ptolemy had to make an assumption that the moon followed a path that sometimes brought it twice as close to the Earth as at other times. And that meant that the moon had sometimes to appear twice as big as it usually does. Ptolemy was aware of this flaw but nevertheless his model was generally, although not universally, accepted. It was adopted by the Christian church as the picture of the universe that was in accordance with Scripture. It had the great advantage that it left lots of room outside the sphere of fixed stars for heaven and hell.

A much simpler model, however, was proposed in 1514 by a Polish priest, Nicholas Copernicus. At first, for fear of being accused of heresy, Copernicus published his model anonymously. His idea was that the sun was stationary at the center and that the Earth and the planets moved in circular orbits around the sun. Sadly for Copernicus, nearly a century passed before this idea was to be taken seriously. Then two astronomers—the German, Johannes Kepler, and the Italian, Galileo Galilei—started publicly to support the Copernican theory, despite the fact that the orbits it predicted did not quite match the ones observed. The death of the Aristotelian-Ptolemaic theory came in 1609. In that year Galileo started observing the night sky with a telescope, which had just been invented.

When he looked at the planet Jupiter, Galileo found that it was accompanied by several small satellites, or moons, which orbited around it. This implied that everything did not have to orbit directly around the Earth as Aristotle and Ptolemy had thought. It was, of course, still possible to believe that the Earth was stationary at the center of the universe, but that the moons of Jupiter moved on extremely complicated paths around the Earth, giving the appearance that they orbited Jupiter. However, Copernicus's theory was much simpler.

At the same time, Kepler had modified Copernicus's theory, suggesting that the planets moved not in circles, but in ellipses. The predictions now finally matched the observations. As far as Kepler was concerned, elliptical orbits were merely an ad hoc hypothesis—and a rather repugnant one at that because ellipses were clearly less perfect than circles. Having discovered, almost by accident, that elliptical orbits fitted the observations well, he could not reconcile with his idea that the planets were made to orbit the sun by magnetic forces.

An explanation was provided only much later, in 1687, when Newton published his *Principia Mathematica Naturalis Causae*. This was probably the most important single work ever published in the physical sciences. In it, Newton not only put forward a theory of how bodies moved in space and time, but he also developed the mathematics needed to analyze those motions. In addition, Newton postulated a law of universal gravitation. This said that each body in the universe was attracted toward every other body by a force which was stronger the more massive the bodies and the closer they were to each other. It was the same force which caused objects to fall to the ground. The story that Newton was hit on the head by an apple is almost certainly apocryphal. All Newton himself ever said was that the idea of gravity came to him as he sat in a contemplative mood, and was occasioned by the fall of an apple.

Newton went on to show that, according to his law, gravity causes the moon to move in an elliptical orbit around the Earth and causes the Earth and the planets to follow elliptical paths around the sun. The Copernican model got rid of Ptolemy's celestial spheres, and with them the idea that the universe had a natural boundary. The fixed stars did not appear to change their relative positions as the Earth went around the sun. It therefore became natural to suppose that the fixed stars were objects like our sun but much farther away. This raised a problem. Newton realized that, according to his theory of gravity, the stars should attract each other; so, it seemed they could not remain essentially motionless. Would they not all fall together at some point?

In a letter in 1691 to Richard Bentley, another leading thinker of his day, Newton argued that this would indeed happen if there were only a finite number of stars. But he reasoned that if, on the other hand, there were an infinite number of stars distributed more or less uniformly over infinite space, this would not happen because there would not be any central point for them to fall to. This argument is an instance of the pitfalls that one can encounter when one talks about infinity.

In an infinite universe, every point can be regarded as the center because every point has an infinite number of stars on each side of it. The correct approach, it was realized only much later, is to consider the finite situation in which the stars all fall in on each other. One then asks how things change if one adds more stars roughly uniformly distributed outside this region. According to Newton's law, the extra stars would make no difference at all to the original ones, and so the stars would fall in just as fast. We can add as many stars as we like, but they will still always collapse in on themselves. We now know it is impossible to have an infinite static model of the universe in which gravity is always attractive.

It is an interesting reflection on the general climate of thought before the twentieth century that no one had suggested that the universe was expanding or contracting. It was generally accepted that either the universe had existed forever in an unchanging state or that it had been created at a finite time in the past, more or less as we observe it today. In part, this may have been due to people's tendency to believe in eternal truths as well as the comfort they found in the thought that even though they may grow old and die, the universe is unchanging.

Even those who realized that Newton's theory of gravity showed that the universe could not be static did not think to suggest that it might be expanding. Instead, they attempted to modify the theory by making the gravitational force repulsive at very large distances. This did not significantly affect their predictions of the motions of the planets. But it would allow an infinite distribution of stars to remain in equilibrium, with the attractive forces between nearby stars being balanced by the repulsive forces from those that were farther away.

However, we now believe such an equilibrium would be unstable. If the stars in some region got only slightly near each other, the attractive forces between them would become stronger and would dominate over the repulsive forces. This would mean that the stars would continue to fall toward each other. On the other hand, if the stars got a bit farther away from each other, the repulsive forces would dominate and drive them farther apart.

Another objection to an infinite static universe is normally ascribed to the German philosopher Heinrich Olbers. In fact, various contemporaries of Newton had raised the problem, and the Olbers article of 1823 was not even the first to contain plausible arguments on this subject. It was, however, the first to be widely noted. The difficulty is that in an infinite static universe nearly every would end on the surface of a star. Thus one would expect that the whole sky would be as bright as the sun, even at night. Olbers's counterargument was that the light from distant stars would be dimmed by absorption by intervening matter. However, if that happened, the intervening matter would eventually heat up until it glowed as brightly as the stars.

The only way of avoiding the conclusion that the whole of the night sky should be as bright as the surface of the sun would be if the stars had not been shining forever, but had turned on at some finite time in the past. In that case, the absorbing matter might not have heated up yet, or the light from distant stars might not yet have reached us. And that brings us to the question of what could have caused the stars to have turned on in the first place.

THE BEGINNING OF THE UNIVERSE

The beginning of the universe had, of course, been discussed for a long time. According to a number of early cosmologies in the Jewish/Christian/Muslim tradition, the universe started at a finite and not very distant time in the past. One argument for such a beginning was the feeling that it was necessary to have a first cause to explain the existence of the universe.

Another argument was put forward by St. Augustine in his book, *The City of God*. He pointed out that civilization is progressing, and we remember who performed this deed or developed that technique. Thus man, and so also perhaps the universe, could not have been around all that long. For otherwise we would have already progressed more than we have.

St. Augustine accepted a date of about 5000 B.C. for the creation of the universe according to the book of Genesis. It is interesting that this is not so far from the end of the last Ice Age, about 10,000 B.C., which is when civilization really began. Aristotle and most of the other Greek philosophers, on the other hand, did not like the idea of a creation because it made too much of divine intervention. They believed, therefore, that the human race and the world around it had existed, and would exist, forever. They had already considered the argument about progress, described earlier, and answered it by saying that there had been periodic floods or other disasters that repeatedly set the human race right back to the beginning of civilization.

When most people believed in an essentially static and unchanging universe, the question of whether or not it had a beginning was really one of metaphysics or theology. One could account for what was observed either way. Either the universe had existed forever, or it was set in motion at some finite time in such a manner as to look as though it had existed forever. But in 1929, Edwin Hubble made the landmark observation that wherever you look, distant stars are moving rapidly away from us. In other words, the universe is expanding. This means that at earlier times objects would have been closer together. In fact, it seemed that there was a time about ten or twenty thousand million years ago when they were all at exactly the same place.

This discovery finally brought the question of the beginning of the universe into the realm of science. Hubble's observations suggested that there was a time called the big bang when the universe was infinitesimally small and, therefore, infinitely dense. If there were events earlier than this time, then they could not affect what happens at the present time. Their existence can be ignored because it would have no observational consequences.

One may say that time had a beginning at the big bang, in the sense that earlier times simply could not be defined. It should be emphasized that this beginning in time is very different from those that had been considered previously. In an unchanging universe, a beginning in time is something that has to be imposed by some being outside the universe. There is no physical necessity for a beginning. One can imagine that God created the universe at literally any time in the past. On the other hand, if the universe is expanding, there may be physical reasons why there had to be a beginning. One could still believe that God created the universe at the instant of the big bang. He could even have created it at a later time in just such a way as to make it look as though there had been a big bang. But it would be meaningless to suppose that it was created before the big bang. An expanding universe does not preclude a creator, but it does place limits on when He might have carried out his job.

LECTURE *2*
THE EXPANDING UNIVERSE

Our sun and the nearby stars are all part of a vast collection of stars called the Milky Way galaxy. For a long time it was thought that this was the whole universe. It was only in 1924 that the American astronomer Edwin Hubble demonstrated that ours was not the only galaxy. There were, in fact, many others, with vast tracks of empty space between them. In order to prove this he needed to determine the distances to these other galaxies. We can determine the distance of nearby stars by observing how they change position as the Earth goes around the sun. But other galaxies are so far away that, unlike nearby stars, they really do appear fixed. Hubble was forced, therefore, to use indirect methods to measure the distances.

Now the apparent brightness of a star depends on two factors—luminosity and how far it is from us. For nearby stars we can measure both their apparent brightness and their distance, so we can work out their luminosity. Conversely, if we knew the luminosity of stars in other galaxies, we could work out their distance by measuring their apparent brightness. Hubble argued that there were certain types of stars that always had the same luminosity when they were near enough for us to measure. If, therefore, we found such stars in another galaxy, we could assume that they had the same luminosity. Thus, we could calculate the distance to that galaxy. If we could do this for a number of stars in the same galaxy, and our calculations always gave the same distance, we could be fairly confident of our estimate. In this way, Edwin Hubble worked out the distances to nine different galaxies.

We now know that our galaxy is only one of some hundred thousand million that can be seen using modern telescopes, each galaxy itself containing some hundred thousand million stars. We live in a galaxy that is about one hundred thousand light-years across and is slowly rotating; the stars in its spiral arms orbit around its center about once every hundred million years. Our sun is just an ordinary, average-sized, yellow star, near the outer edge of one of the spiral arms. We have certainly come a long way since Aristotle and Ptolemy, when we thought that the Earth was the center of the universe.

Stars are so far away that they appear to us to be just pinpoints of light. We cannot determine their size or shape. So how can we tell different types of stars apart? For the vast majority of stars, there is only one correct characteristic feature that we can observe—the color of their light. Newton discovered that if light from the sun passes through a prism, it breaks up into its component colors—its spectrum—like in a rainbow. By focusing a telescope on an individual star or galaxy, one can similarly observe the spectrum of the light from that star or galaxy. Different stars have different spectra, but the relative brightness of the different colors is always exactly what one would expect to find in the light emitted by an object that is glowing red hot. This means that we can tell a star's temperature from the spectrum of its light. Moreover, we find that certain very specific colors are missing from stars' spectra, and these missing colors may vary from star to star. We know that each chemical element absorbs the characteristic set of very specific colors. Thus, by matching each of those which are missing from a star's spectrum, we can determine exactly which elements are present in the star's atmosphere.

In the 1920s, when astronomers began to look at the spectra of stars in other galaxies, they found something most peculiar: There were the same characteristic sets of missing colors as for stars in our own galaxy, but they were all shifted by the same relative amount toward the red end of the spectrum. The only reasonable explanation of this was that the galaxies were moving away from us, and the frequency of the light waves from them was being reduced, or red-shifted, by the Doppler effect. Listen to a car passing on the road. As the car is approaching, its engine sounds at a higher pitch, corresponding to a higher frequency of sound waves; and when it passes and goes away, it sounds at a lower pitch. The behavior of light or radial waves is similar. Indeed, the police made use of the Doppler effect to measure the speed of cars by measuring the frequency of pulses of radio waves reflected off them.

In the years following his proof of the existence of other galaxies, Hubble spent his time cataloging their distances and observing their spectra. At that time most people expected the galaxies to be moving around quite randomly, and so expected to find as many spectra which were blue-shifted as ones which were red-shifted. It was quite a surprise, therefore, to find that the galaxies all appeared red-shifted. Every single one was moving away from us. More surprising still was the result which Hubble published in 1929: Even the size of the galaxy's red shift was not random, but was directly proportional to the galaxy's distance from us. Or, in other words, the farther a galaxy was, the faster it was moving away. And that meant that the universe could not be static, as everyone previously thought, but was in fact expanding. The distance between the different galaxies was growing all the time.

The discovery that the universe was expanding was one of the great intellectual revolutions of the twentieth century. With hindsight, it is easy to wonder why no one had thought of it before. Newton and others should have realized that a static universe would soon start to contract under the influence of gravity. But suppose that, instead of being static, the universe was expanding. If it was expanding fairly slowly, the force of gravity would cause it eventually to stop expanding and then to start contracting. However, if it was expanding at more than a certain critical rate, gravity would never be strong enough to stop it, and the universe would continue to expand forever. This is a bit like what happens when one fires a rocket upward from the surface of the Earth. If it has a fairly low speed, gravity will eventually stop the rocket and it will start falling back. On the other hand, if the rocket has more than a certain critical speed-about seven miles a second-gravity will not be strong enough to pull it back, so it will keep going away from the Earth forever.

This behavior of the universe could have been predicted from Newton's theory of gravity at any time in the nineteenth, the eighteenth, or even the late seventeenth centuries. Yet so strong was the belief in a static universe that it persisted into the early twentieth century. Even when Einstein formulated the general theory of relativity in 1915, he was sure that the universe had to be static. He therefore modified his theory to make this possible, introducing a socalled cosmological constant into his equations. This was a new "antigravity" force, which, unlike other forces, did not come from any particular source, but was built into the very fabric of space-time. His cosmological constant gave space-time an inbuilt tendency to expand, and this could be made to exactly balance the attraction of all the matter in the universe so that a static universe would result.

Only one man, it seems, was willing to take general relativity at face value. While Einstein and other physicists were looking for ways of avoiding general relativity's prediction of a nonstatic universe, the Russian physicist Alexander Friedmann instead set about explaining it.

THE FRIEDMANN MODELS

The equations of general relativity, which determined how the universe evolves in time, are too complicated to solve in detail. So what Friedmann did, instead, was to make two very simple assumptions about the universe: that the universe looks identical in whichever direction we look, and that this would also be true if we were observing the universe from anywhere else. On the basis of general relativity and these two assumptions, Friedmann showed that we should not expect the universe to be static. In fact, in 1922, several years before Edwin Hubble's discovery, Friedmann predicted exactly what Hubble found.

The assumption that the universe looks the same in every direction is clearly not true in reality. For example, the other stars in our galaxy form a distinct band of light across the night sky called the Milky Way. But if we look at distant galaxies, there seems to be more or less the same number of them in each direction. So the universe does seem to be roughly the same in every direction, provided one views it on a large scale compared to the distance between galaxies.

For a long time this was sufficient justification for Friedmann's assumption—as a rough approximation to the real universe. But more recently a lucky accident uncovered the fact that Friedmann's assumption is in fact a remarkably accurate description of our universe. In 1965, two American physicists, Arno Penzias and Robert Wilson, were working at the Bell Labs in New Jersey on the design of a very sensitive microwave detector for communicating with orbiting satellites. They were worried when they found that their detector was picking up more noise than it ought to, and that the noise did not appear to be coming from any particular direction. First they looked for bird droppings on their detector and checked for other possible malfunctions, but soon ruled these out. They knew that any noise from within the atmosphere would be stronger when the detector is not pointing straight up than when it is, because the atmosphere appears thicker when looking at an angle to the vertical.

The extra noise was the same whichever direction the detector pointed, so it must have come from outside the atmosphere. It was also the same day and night throughout the year, even though the Earth was rotating on its axis and orbiting around the sun. This showed that the radiation must come from beyond the solar system, and even from beyond the galaxy, as otherwise it would vary as the Earth pointed the detector in different directions.

In fact, we know that the radiation must have traveled to us across most of the observable universe. Since it appears to be the same in different directions, the universe must also be the same in every direction, at least on a large scale. We now know that whichever direction we look in, this noise never varies by more than one part in ten thousand. So Penzias and Wilson had unwittingly stumbled across a remarkably accurate confirmation of Friedmann's first assumption.

At roughly the same time, two American physicists at nearby Princeton University, Bob Dicke and Jim Peebles, were also taking an interest in microwaves. They were working on a suggestion made by George Gamow, once a student of Alexander Friedmann, that the early universe should have been very hot and dense, glowing white hot. Dicke and Peebles argued that we should still be able to see this glowing, because light from very distant parts of the early universe would only just be reaching us now. However, the expansion of the universe meant that this light should be so greatly red-shifted that it would appear to us now as microwave radiation. Dicke and Peebles were looking for this radiation when Penzias and Wilson heard about their work and realized that they had already found it. For this, Penzias and Wilson were awarded the Nobel Prize in 1978, which seems a bit hard on Dicke and Peebles.

Now at first sight, all this evidence that the universe looks the same whichever direction we look in might seem to suggest there is something special about our place in the universe. In particular, it might seem that if we observe all other galaxies to be moving away from us, then we must be at the center of the universe. There is, however, an alternative explanation: The universe might also look the same in every direction as seen from any other galaxy. This, as we have seen, was Friedmann's second assumption.

We have no scientific evidence for or against this assumption. We believe it only on grounds of modesty. It would be most remarkable if the universe looked the same in every direction around us, but not around other points in the universe. In Friedmann's model, all the galaxies are moving directly away from each other. The situation is rather like steadily blowing up a balloon which has a number of spots painted on it. As the balloon expands, the distance between any two spots increases, but there is no spot that can be said to be the center of the expansion. Moreover, the farther apart the spots are, the faster they will be moving apart. Similarly, in Friedmann's model the speed at which any two galaxies are moving apart is proportional to the distance between them. So it predicted that the red shift of a galaxy should be directly proportional to its distance from us, exactly as Hubble found.

Despite the success of his model and his prediction of Hubble's observations, Friedmann's work remained largely unknown in the West. It became known only after similar models were discovered in 1935 by the American physicist Howard Robertson and the British mathematician Arthur Walker, in response to Hubble's discovery of the uniform expansion of the universe.

Although Friedmann found only one, there are in fact three different kinds of models that obey Friedmann's two fundamental assumptions. In the first kind—which Friedmann found—the universe is expanding so sufficiently slowly that the gravitational attraction be tween the different galaxies causes the expansion to slow down and eventually to stop. The galaxies then start to move toward each other and the universe contracts. The distance between two neighboring galaxies starts at zero, increases to a maximum, and then decreases back down to zero again.

In the second kind of solution, the universe is expanding so rapidly that the gravitational attraction can never stop it, though it does slow it down a bit. The separation between neighboring galaxies in this model starts at zero, and eventually the galaxies are moving apart at a steady speed.

Finally, there is a third kind of solution, in which the universe is expanding only just fast enough to avoid recollapse. In this case the separation also starts at zero, and increases forever. However, the speed at which the galaxies are moving apart gets smaller and smaller, although it never quite reaches zero.

A remarkable feature of the first kind of Friedmann model is that the universe is not infinite in space, but neither does space have any boundary. Gravity is so strong that space is bent round onto itself, making it rather like the surface of the Earth. If one keeps traveling in a certain direction on the surface of the Earth, one never comes up against an impassable barrier or falls over the edge, but eventually comes back to where one started. Space, in the first Friedmann model, is just like this, but with three dimensions instead of two for the Earth's surface. The fourth dimension—time—is also finite in extent, but it is like a line with two ends or boundaries, a beginning and an end. We shall see later that when one combines general relativity with the uncertainty principle of quantum mechanics, it is possible for both space and time to be finite without any edges or boundaries. The idea that one could go right around the universe and end up where one started makes good science fiction, but it doesn't have much practical significance because it can be shown that the universe would recollapse to zero size before one could get round. You would need to travel faster than light in order to end up where you started before the universe came to an end—and that is not allowed.

But which Friedmann model describes our universe? Will the universe eventually stop expanding and start contracting, or will it expand forever? To answer this question we need to know the present rate of expansion of the universe and its present average density. If the density is less than a certain critical value, determined by the rate of expansion, the gravitational attraction will be too weak to halt the expansion. If the density is greater than the critical value, gravity will stop the expansion at some time in the future and cause the universe to recollapse.

We can determine the present rate of expansion by measuring the velocities at which other galaxies are moving away from us, using the Doppler effect. This can be done very accurately. However, the distances to the galaxies are not very well known because we can only measure them indirectly. So all we know is that the universe is expanding by between 5 percent and 10 percent every thousand million years. However, our uncertainty about the present average density of the universe is even greater.

If we add up the masses of all the stars that we can see in our galaxy and other galaxies, the total is less than one-hundredth of the amount required to halt the expansion of the universe, even in the lowest estimate of the rate of expansion. But we know that our galaxy and other galaxies must contain a large amount of dark matter which we cannot see directly, but which we know must be there because of the influence of its gravitational attraction on the orbits of stars and gas in the galaxies. Moreover, most galaxies are found in clusters, and we can similarly infer the presence of yet more dark matter in between the galaxies in these clusters by its effect on the motion of the galaxies. When we add up all this dark matter, we still get only about one-tenth of the amount required to halt the expansion. However, there might be some other form of matter which we have not yet detected and which might still raise the average density of the universe up to the critical value needed to halt the expansion.

The present evidence, therefore, suggests that the universe will probably expand forever. But don't bank on it. All we can really be sure of is that even if the universe is going to recollapse, it won't do so for at least another ten thousand million years, since it has already been expanding for at least that long. This should not unduly worry us since by that time, unless we have colonies beyond the solar system, mankind will long since have died out, extinguished along with the death of our sun.

THE BIG BANG

All of the Friedmann solutions have the feature that at some time in the past, between ten and twenty thousand million years ago, the distance between neighboring galaxies must have been zero. At that time, which we call the big bang, the density of the universe and the curvature of space-time would have been infinite. This means that the general theory of relativity—on which Friedmann's solutions are based—predicts that there is a singular point in the universe.

All our theories of science are formulated on the assumption that space-time is smooth and nearly flat, so they would all break down at the big bang singularity, where the curvature of space-time is infinite. This means that even if there were events before the big bang, one could not use them to determine what would happen afterward, because predictability would break down at the big bang. Correspondingly, if we know only what has happened since the big bang, we could not determine what happened beforehand. As far as we are concerned, events before the big bang can have no consequences, so they should not form part of a scientific model of the universe. We should therefore cut them out of the model and say that time had a beginning at the big bang.

Many people do not like the idea that time has a beginning, probably because it smacks of divine intervention. (The Catholic church, on the other hand, had seized on the big bang model and in 1951 officially pronounced it to be in accordance with the Bible.) There were a number of attempts to avoid the conclusion that there had been a big bang. The proposal that gained widest support was called the steady state theory. It was suggested in 1948 by two refugees from Nazi-occupied Austria, Hermann Bondi and Thomas Gold, together with the Briton Fred Hoyle, who had worked with them on the development of radar during the war. The idea was that as the galaxies moved away from each other, new galaxies were continually forming in the gaps in between, from new matter that was being continually created. The universe would therefore look roughly the same at all times as well as at all points of space.

The steady state theory required a modification of general relativity to allow for the continual creation of matter, but the rate that was involved was so low—about one particle per cubic kilometer per year—that it was not in conflict with experiment. The theory was a good scientific theory, in the sense that it was simple and it made definite predictions that could be tested by observation. One of these predictions was that the number of galaxies or similar objects in any given volume of space should be the same wherever and whenever we look in the universe.

In the late 1950s and early 1960s, a survey of sources of radio waves from outer space was carried out at Cambridge by a group of astronomers led by Martin Ryle. The Cambridge group showed that most of these radio sources must lie outside our galaxy, and also that there were many more weak sources than strong ones. They interpreted the weak sources as being the more distant ones, and the stronger ones as being near. Then there appeared to be fewer sources per unit volume of space for the nearby sources than for the distant ones.

This could have meant that we were at the center of a great region in the universe in which the sources were fewer than elsewhere. Alternatively, it could have meant that the sources were more numerous in the past, at the time that the radio waves left on their journey to us, than they are now. Either explanation contradicted the predictions of the steady state theory. Moreover, the discovery of the microwave radiation by Penzias and Wilson in 1965 also indicated that the universe must have been much denser in the past. The steady state theory therefore had regretfully to be abandoned.

Another attempt to avoid the conclusion that there must have been a big bang and, therefore, a beginning of time, was made by two Russian scientists, Evgenii Lifshitz and Isaac Khalatnikov, in 1963. They suggested that the big bang might be a peculiarity of Friedmann's models alone, which after all were only approximations to the real universe. Perhaps, of all the models that were roughly like the real universe, only Friedmann's would contain a big bang singularity. In Friedmann's models, the galaxies are all moving directly away from each other. So it is not surprising that at some time in the past they were all at the same place. In the real universe, however, the galaxies are not just moving directly away from each other—they also have small sideways velocities. So in reality they need never have been all at exactly the same place, only very close together. Perhaps, then, the current expanding universe resulted not from a big bang singularity, but from an earlier contracting phase; as the universe had collapsed, the particles in it might not have all collided, but they might have flown past and then away from each other, producing the present expansion of the universe. How then could we tell whether the real universe should have started out with a big bang?

What Lifshitz and Khalatnikov did was to study models of the universe which were roughly like Friedmann's models but which took account of the irregularities and random velocities of galaxies in the real universe. They showed that such models could start with a big bang, even though the galaxies were no longer always moving directly away from each other. But they claimed that this was still only possible in certain exceptional models in which the galaxies were all moving in just the right way. They argued that since there seemed to be infinitely more Friedmann-like models without a big bang singularity than there were with one, we should conclude that it was very unlikely that there had been a big bang. They later realized, however, that there was a much more general class of Friedmann-like models which did have singularities, and in which the galaxies did not have to be moving in any special way. They therefore withdrew their claim in 1970.

The work of Lifshitz and Khalatnikov was valuable because it showed that the universe could have had a singularity—a big bang—if the general theory of relativity was correct. However, it did not resolve the crucial question: Does general relativity predict that our universe should have the big bang, a beginning of time? The answer to this came out of a completely different approach started by a British physicist, Roger Penrose, in 1965. He used the way light cones behave in general relativity, and the fact that gravity is always attractive, to show that a star that collapses under its own gravity is trapped in a region whose boundary eventually shrinks to zero size. This means that all the matter in the star will be compressed into a region of zero volume, so the density of matter and the curvature of space-time become infinite. In other words, one has a singularity contained within a region of space-time known as a black hole.

At first sight, Penrose's result didn't have anything to say about the question of whether there was a big bang singularity in the past.

However, at the time that Penrose produced his theorem, I was a research student desperately looking for a problem with which to complete my Ph.D. thesis. I realized that if one reversed the direction of time in Penrose's theorem so that the collapse became an expansion, the conditions of his theorem would still hold, provided the universe were roughly like a Friedmann model on large scales at the present time. Penrose's theorem had shown that any collapsing star must end in a singularity; the time-reversed argument showed that any Friedmann-like expanding universe must have begun with a singularity. For technical reasons, Penrose's theorem required that the universe be infinite in space. So I could use it to prove that there should be a singularity only if the universe was expanding fast enough to avoid collapsing again, because only that Friedmann model was infinite in space.

During the next few years I developed new mathematical techniques to remove this and other technical conditions from the theorems that proved that singularities must occur. The final result was a joint paper by Penrose and myself in 1970, which proved that there must have been a big bang singularity provided only that general relativity is correct and that the universe contains as much matter as we observe.

There was a lot of opposition to our work, partly from the Russians, who followed the party line laid down by Lifshitz and Khalatnikov, and partly from people who felt that the whole idea of singularities was repugnant and spoiled the beauty of Einstein's theory. However, one cannot really argue with the mathematical theorem. So it is now generally accepted that the universe must have a beginning.

LECTURE 3
BLACK HOLES

The term *black hole* is of very recent origin. It was coined in 1969 by the American scientist John Wheeler as a graphic description of an idea that goes back at least two hundred years. At that time there were two theories about light. One was that it was composed of particles; the other was that it was made of waves. We now know that really both theories are correct. By the wave/particle duality of quantum mechanics, light can be regarded as both a wave and a particle. Under the theory that light was made up of waves, it was not clear how it would respond to gravity. But if light were composed of particles, one might expect them to be affected by gravity in the same way that cannonballs, rockets, and planets are.

On this assumption, a Cambridge don, John Michell, wrote a paper in 1783 in the *Philosophical Transactions of the Royal Society of London*. In it, he pointed out that a star that was sufficiently massive and compact would have such a strong gravitational field that light could not escape. Any light emitted from the surface of the star would be dragged back by the star's gravitational attraction before it could get very far. Michell suggested that there might be a large number of stars like this. Although we would not be able to see them because the light from them would not reach us, we would still feel their gravitational attraction. Such objects are what we now call black holes, because that is what they are—black voids in space.

A similar suggestion was made a few years later by the French scientist the Marquis de Laplace, apparently independently of

Michell. Interestingly enough, he included it in only the first and second editions of his book, The *System of the World*, and left it out of later editions; perhaps he decided that it was a crazy idea. In fact, it is not really consistent to treat light like cannonballs in Newton's theory of gravity because the speed of light is fixed. A cannonball fired upward from the Earth will be slowed down by gravity and will eventually stop and fall back. A photon, however, must continue upward at a constant speed. How, then, can Newtonian gravity affect light? A consistent theory of how gravity affects light did not come until Einstein proposed general relativity in 1915; and even then it was a long time before the implications of the theory for massive stars were worked out.

To understand how a black hole might be formed, we first need an understanding of the life cycle of a star. A star is formed when a large amount of gas, mostly hydrogen, starts to collapse in on itself due to its gravitational attraction. As it contracts, the atoms of the gas collide with each other more and more frequently and at greater and greater speeds—the gas heats up. Eventually the gas will be so hot that when the hydrogen atoms collide they no longer bounce off each other but instead merge with each other to form helium atoms. The heat released in this reaction, which is like a controlled hydrogen bomb, is what makes the stars shine. This additional heat also increases the pressure of the gas until it is sufficient to balance the gravitational attraction, and the gas stops contracting. It is a bit like a balloon where there is a balance between the pressure of the air inside, which is trying to make the balloon expand, and the tension in the rubber, which is trying to make the balloon smaller.

The stars will remain stable like this for a long time, with the heat from the nuclear reactions balancing the gravitational attraction. Eventually, however, the star will run out of its hydrogen and other nuclear fuels. And paradoxically, the more fuel a star starts off with, the sooner it runs out. This is because the more massive the star is, the hotter it needs to be to balance its gravitational attraction. And the hotter it is, the faster it will use up its fuel. Our sun has probably got enough fuel for another five thousand million years or so, but more massive stars can use up their fuel in as little as one hundred million years, much less than the age of the universe. When the star runs out of fuel, it will start to cool off and so to contract. What might happen to it then was only first understood at the end of the 1920s.

In 1928 an Indian graduate student named Subrahmanyan Chandrasekhar set sail for England to study at Cambridge with the British astronomer Sir Arthur Eddington. Eddington was an expert on general relativity. There is a story that a journalist told Eddington in the early 1920s that he had heard there were only three people in the world who understood general relativity. Eddington replied, "I am trying to think who the third person is."

During his voyage from India, Chandrasekhar worked out how big a star could be and still separate itself against its own gravity after it had used up all its fuel. The idea was this: When the star becomes small, the matter particles get very near each other. But the Pauli exclusion principle says that two matter particles cannot have both the same position and the same velocity. The matter particles must therefore have very different velocities. This makes them move away from each other, and so tends to make the star expand. A star can therefore maintain itself at a constant radius by a balance between the attraction of gravity and the repulsion that arises from the exclusion principle, just as earlier in its life the gravity was balanced by the heat.

Chandrasekhar realized, however, that there is a limit to the repulsion that the exclusion principle can provide. The theory of relativity limits the maximum difference in the velocities of the matter particles in the star to the speed of light. This meant that when the star got sufficiently dense, the repulsion caused by the exclusion principle would be less than the attraction of gravity. Chandrasekhar calculated that a cold star of more than about one and a half times the mass of the sun would not be able to support itself against its own gravity. This mass is now known as the *Chandrasekhar limit*.

This had serious implications for the ultimate fate of massive stars. If a star's mass is less than the Chandrasekhar limit, it can eventually stop contracting and settle down to a possible final state as a *white dwarf* with a radius of a few thousand miles and a density of hundreds of tons per cubic inch. A white dwarf is supported by the exclusion principle repulsion between the electrons in its matter. We observe a large number of these white dwarf stars. One of the first to be discovered is the star that is orbiting around Sirius, the brightest star in the night sky.

It was also realized that there was another possible final state for a star also with a limiting mass of about one or two times the mass of the sun, but much smaller than even the white dwarf. These stars would be supported by the exclusion principle repulsion between the neutrons and protons, rather than between the electrons. They were therefore called neutron stars. They would have had a radius of only ten miles or so and a density of hundreds of millions of tons per cubic inch. At the time they were first predicted, there was no way that neutron stars could have been observed, and they were not detected until much later.

Stars with masses above the Chandrasekhar limit, on the other hand, have a big problem when they come to the end of their fuel. In some cases they may explode or manage to throw off enough matter to reduce their mass below the limit, but it was difficult to believe that this always happened, no matter how big the star. How would it know that it had to lose weight? And even if every star managed to lose enough mass, what would happen if you added more mass to a white dwarf or neutron star to take it over the limit? Would it collapse to infinite density?

Eddington was shocked by the implications of this and refused to believe Chandrasekhar's result. He thought it was simply not possible that a star could collapse to a point. This was the view of most scientists. Einstein himself wrote a paper in which he claimed that stars would not shrink to zero size.The hostility of other scientists, particularly of Eddington, his former teacher and the leading authority on the structure of stars, persuaded Chandrasekhar to abandon this line of work and turn instead to other problems in astronomy. However, when he was awarded the Nobel Prize in 1983, it was, at least in part, for his early work on the limiting mass of cold stars.

Chandrasekhar had shown that the exclusion principle could not halt the collapse of a star more massive than the Chandrasekhar limit. But the problem of understanding what would happen to such a star, according to general relativity, was not solved until 1939 by a young American, Robert Oppenheimer. His result, however, suggested that there would be no observational consequences that could be detected by the telescopes of the day. Then the war intervened and Oppenheimer himself became closely involved in the atom bomb project. And after the war the problem of gravitational collapse was largely forgotten as most scientists were then interested in what happens on the scale of the atom and its nucleus. In the 1960s, however, interest in the large-scale problems of astronomy and cosmology was revived by a great increase in the number and range of astronomical observations brought about by the application of modern technology. Oppenheimer's work was then rediscovered and extended by a number of people.

The picture that we now have from Oppenheimer's work is as follows: The gravitational field of the star changes the paths of light rays in space-time from what they would have been had the star not been present. The light cones, which indicate the paths followed in space and time by flashes of light emitted from their tips, are bent slightly inward near the surface of the star. This can be seen in the bending of light from distant stars that is observed during an eclipse of the sun. As the star contracts, the gravitational field at its surface gets stronger and the light cones get bent inward more. This makes it more difficult for light from the star to escape, and the light appears dimmer and redder to an observer at a distance.

Eventually, when the star has shrunk to a certain critical radius, the gravitational field at the surface becomes so strong that the light cones are bent inward so much that the light can no longer escape. According to the theory of relativity, nothing can travel faster than light. Thus, if light cannot escape, neither can anything else. Everything is dragged back by the gravitational field. So one has a set of events, a region of space-time, from which it is not possible to escape to reach a distant observer. This region is what we now call a black hole. Its boundary is called the event horizon. It coincides with the paths of the light rays that just fail to escape from the black hole.

In order to understand what you would see if you were watching a star collapse to form a black hole, one has to remember that in the theory of relativity there is no absolute time. Each observer has his own measure of time. The time for someone on a star will be different from that for someone at a distance, because of the gravitational field of the star. This effect has been measured in an experiment on Earth with clocks at the top and bottom of a water tower. Suppose an intrepid astronaut on the surface of the collapsing star sent a signal every second, according to his watch, to his spaceship orbiting about the star. At some time on his watch, say eleven o'clock, the star would shrink below the critical radius at which the gravitational field became so strong that the signals would no longer reach the spaceship.

His companions watching from the spaceship would find the intervals between successive signals from the astronaut getting longer and longer as eleven o'clock approached. However, the effect would be very small before 10:59:59. They would have to wait only very slightly more than a second between the astronaut's 10:59:58 signal and the one that he sent when his watch read 10:59:59, but they would have to wait forever for the eleven o'clock signal. The light waves emitted from the surface of the star between 10:59:59 and eleven o'clock, by the astronaut's watch, would be spread out over an infinite period of time, as seen from the spaceship.

The time interval between the arrival of successive waves at the spaceship would get longer and longer, and so the light from the star would appear redder and redder and fainter and fainter. Eventually the star would be so dim that it could no longer be seen from the spaceship. All that would be left would be a black hole in space. The star would, however, continue to exert the same gravitational force on the spaceship. This is because the star is still visible to the spaceship, at least in principle. It is just that the light from the surface is so red-shifted by the gravitational field of the star that it cannot be seen. However, the red shift does not affect the gravitational field of the star itself. Thus, the spaceship would continue to orbit the black hole.

The work that Roger Penrose and I did between 1965 and 1970 showed that, according to general relativity, there must be a singularity of infinite density within the black hole. This is rather like the big bang at the beginning of time, only it would be an end of time for the collapsing body and the astronaut. At the singularity, the laws of science and our ability to predict the future would break down. However, any observer who remained outside the black hole would not be affected by this failure of predictability, because neither light nor any other signal can reach them from the singularity.

This remarkable fact led Roger Penrose to propose the cosmic censorship hypothesis, which might be paraphrased as "God abhors a naked singularity." In other words, the singularities produced by gravitational collapse occur only in places like black holes, where they are decently hidden from outside view by an event horizon.

Strictly, this is what is known as the weak cosmic censorship hypothesis: protect obervers who remain outside the black hole from the consequences of the breakdown of predictability that occurs at the singularity. But it does nothing at all for the poor unfortunate astronaut who falls into the hole. Shouldn't God protect his modesty as well?

There are some solutions of the equations of general relativity in which it is possible for our astronaut to see a naked singularity. He may be able to avoid hitting the singularity and instead fall through a "worm hole" and come out in another region of the universe. This would offer great possibilities for travel in space and time, but unfortunately it seems that the solutions may all be highly unstable. The least disturbance, such as the presence of an astronaut, may change them so that the astronaut cannot see the singularity until he hits it and his time comes to an end. In other words, the singularity always lies in his future and never in his past.

The strong version of the cosmic censorship hypothesis states that in a realistic solution, the singularities always lie either entirely in the future, like the singularities of gravitational collapse, or entirely in the past, like the big bang. It is greatly to be hoped that some version of the censorship hypothesis holds, because close to naked singularities it may be possible to travel into the past. While this would be fine for writers of science fiction, it would mean that no one's life would ever be safe. Someone might go into the past and kill your father or mother before you were conceived.

In a gravitational collapse to form a black hole, the movements would be dammed by the emission of gravitational waves. One would therefore expect that it would not be too long before the black hole would settle down to a stationary state. It was generally supposed that this final stationary state would depend on the details of the body that had collapsed to form the black hole. The black hole might have any shape or size, and its shape might not even be fixed, but instead be pulsating.

However, in 1967, the study of black holes was revolutionized by a paper written in Dublin by Werner Israel. Israel showed that any black hole that is not rotating must be perfectly round or spherical. Its size, moreover, would depend only on its mass. It could, in fact, be described by a particular solution of Einstein's equations that had been known since 1917, when it had been found by Karl Schwarzschild shortly after the discovery of general relativity. At first, Israel's result was interpreted by many people, including Israel himself, as evidence that black holes would form only from the collapse of bodies that were perfectly round or spherical. As no real body would be perfectly spherical, this meant that, in general, gravitational collapse would lead to naked singularities.There was, however, a different interpretation of Israel's result, which was advocated by Roger Penrose and John Wheeler in particular. This was that a black hole should behave like a ball of fluid. Although a body might start off in an unspherical state, as it collapsed to form a black hole it would settle down to a spherical state due to the emission of gravitational waves. Further calculations supported this view and it came to be adopted generally.

Israel's result had dealt only with the case of black holes formed from nonrotating bodies. On the analogy with a ball of fluid, one would expect that a black hole made by the collapse of a rotating body would not be perfectly round. It would have a bulge round the equator caused by the effect of the rotation. We observe a small bulge like this in the sun, caused by its rotation once every twenty-five days or so. In 1963, Roy Kerr, a New Zealander, had found a set of black-hole solutions of the equations of general relativity more general than the Schwarzschild solutions. These "Kerr" black holes rotate at a constant rate, their size and shape depending only on their mass and rate of rotation. If the rotation was zero, the black hole was perfectly round and the solution was identical to the

Schwarzschild solution. But if the rotation was nonzero, the black hole bulged outward near its equator. It was therefore natural to conjecture that a rotating body collapsing to form a black hole would end up in a state described by the Kerr solution.

In 1970, a colleague and fellow research student of mine, Brandon Carter, took the first step toward proving this conjecture. He showed that, provided a stationary rotating black hole had an axis of symmetry, like a spinning top, its size and shape would depend only on its mass and rate of rotation. Then, in 1971, I proved that any stationary rotating black hole would indeed have such an axis of symmetry. Finally, in 1973, David Robinson at Kings College, London, used Carter's and my results to show that the conjecture had been correct: Such a black hole had indeed to be the Kerr solution.

So after gravitational collapse a black hole must settle down into a state in which it could be rotating, but not pulsating. Moreover, its size and shape would depend only on its mass and rate of rotation, and not on the nature of the body that had collapsed to form it. This result became known by the maxim "A black hole has no hair." It means that a very large amount of information about the body that has collapsed must be lost when a black hole is formed, because afterward all we can possibly measure about the body is its mass and rate of rotation. The significance of this will be seen in the next lecture. The no-hair theorem is also of great practical importance because it so greatly restricts the possible types of black holes. One can therefore make detailed models of objects that might contain black holes, and compare the predictions of the models with observations.

Black holes are one of only a fairly small number of cases in the history of science where a theory was developed in great detail as a mathematical model before there was any evidence from observations that it was correct. Indeed, this used to be the main argument of opponents of black holes. How could one believe in objects for which the only evidence was calculations based on the dubious theory of general relativity?

In 1963, however, Maarten Schmidt, an astronomer at the Mount Palomar Observatory in California, found a faint, starlike object in the direction of the source of radio waves called 3C273—that is, source number 273 in the third Cambridge catalog of radio sources. When he measured the red shift of the object, he found it was too large to be caused by a gravitational field: If it had been a gravitational red shift, the object would have to be so massive and so near to us that it would disturb the orbits of planets in the solar system. This suggested that the red shift was instead caused by the expansion of the universe, which in turn meant that the object was a very long way away. And to be visible at such a great distance, the object must be very bright and must be emitting a huge amount of energy.

The only mechanism people could think of that would produce such large quantities of energy seemed to be the gravitational collapse not just of a star but of the whole central region of a galaxy. A number of other similar "quasi-stellar objects," or quasars, have since been discovered, all with large red shifts. But they are all too far away, and too difficult, to observe to provide conclusive evidence of black holes.

Further encouragement for the existence of black holes came in 1967 with the discovery by a research student at Cambridge, Jocelyn Bell, of some objects in the sky that were emitting regular pulses of radio waves. At first, Jocelyn and her supervisor, Anthony Hewish, thought that maybe they had made contact with an alien civilization in the galaxy. Indeed, at the seminar at which they announced their discovery, I remember that they called the first four sources to be found LGM 1—4, LGM standing for "Little Green Men."

In the end, however, they and everyone else came to the less romantic conclusion that these objects, which were given the name pulsars, were in fact just rotating neutron stars. They were emitting pulses of radio waves because of a complicated indirection between their magnetic fields and surrounding matter. This was bad news for writers of space westerns, but very hopeful for the small number of us who believed in black holes at that time. It was the first positive evidence that neutron stars existed. A neutron star has a radius of about ten miles, only a few times the critical radius at which a star becomes a black hole. If a star could collapse to such a small size, it was not unreasonable to expect that other stars could collapse to even smaller size and become black holes.

How could we hope to detect a black hole, as by its very definition it does not emit any light? It might seem a bit like looking for a black cat in a coal cellar. Fortunately, there is a way, since as John Michell pointed out in his pioneering paper in 1783, a black hole still exerts a gravitational force on nearby objects. Astronomers have observed a number of systems in which two stars orbit around each other, attracted toward each other by gravity. They also observed systems in which there is only one visible star that is orbiting around some unseen companion.

One cannot, of course, immediately conclude that the companion is a black hole. It might merely be a star that is too faint to be seen. However, some of these systems, like the one called Cygnus X-I, are also strong sources of X rays. The best explanation for this phenomenon is that the X rays are generated by matter that has been blown off the surface of the visible star. As it falls toward the unseen companion, it develops a spiral motion—rather like water running out of a bath—and it gets very hot, emitting X rays. For this mechanism to work, the unseen object has to be very small, like a white dwarf, neutron star, or black hole.

Now, from the observed motion of the visible star, one can determine the lowest possible mass of the unseen object. In the case of Cygnus X-I, this is about six times the mass of the sun. According to Chandrasekhar's result, this is too much for the unseen object to be a white dwarf. It is also too large a mass to be a neutron star. It seems, therefore, that it must be a black hole.

There are other models to explain Cygnus X-I that do not in clude a black hole, but they are all rather far-fetched. A black hole seems to be the only really natural explanation of the observations. Despite this, I have a bet with Kip Thorne of the California Institute of Technology that in fact Cygnus X-I does not contain a black hole. This is a form of insurance policy for me. I have done a lot of work on black holes, and it would all be wasted if it turned out that black holes do not exist. But in that case, I would have the consolation of winning my bet, which would bring me four years of the magazine *Private Eye*. If black holes do exist, Kip will get only one year of *Penthouse*, because when we made the bet in 1975, we were 80 percent certain that Cygnus was a black hole. By now I would say that we are about 95 percent certain, but the bet has yet to be settled.

There is evidence for black holes in a number of other systems in our galaxy, and for much larger black holes at the centers of other galaxies and quasars. One can also consider the possibility that there might be black holes with masses much less than that of the sun. Such black holes could not be formed by gravitational collapse, because their masses are below the Chandrasekhar mass limit. Stars of this low mass can support themselves against the force of gravity even when they have exhausted their nuclear fuel. So, low-mass black holes could form only if matter were compressed to enormous densities by very large external pressures. Such conditions could occur in a very big hydrogen bomb. The physicist John Wheeler once calculated that if one took all the heavy water in all the oceans of the world, one could build a hydrogen bomb that would compress matter at the center so much that a black hole would be created. Unfor-tunately, however, there would be no one left to observe it.

A more practical possibility is that such low-mass black holes might have been formed in the high temperatures and pressures of the very early universe. Black holes could have been formed if the early universe had not been perfectly smooth and uniform, because then a small region that was denser than average could be compressed in this way to form a black hole. But we know that there must have been some irregularities, because otherwise the matter in the universe would still be perfectly uniformly distributed at the present epoch, instead of being clumped together in stars and galaxies.

Whether or not the irregularities required to account for stars and galaxies would have led to the formation of a significant number of these primordial black holes depends on the details of the conditions in the early universe. So if we could determine how many primordial black holes there are now, we would learn a lot about the very early stages of the universe. Primordial black holes with masses more than a thousand million tons—the mass of a large mountain—could be detected only by their gravitational influence on other visible matter or on the expansion of the universe. However, as we shall learn in the next lecture, black holes are not really black after all: They glow like a hot body, and the smaller they are, the more they glow. So, paradoxically, smaller black holes might actually turn out to be easier to detect than large ones.

LECTURE *4*

BLACK HOLES AIN'T SO BLACK

Before 1970, my research on general relativity had concentrated mainly on the question of whether there had been a big bang singularity. However, one evening in November of that year, shortly after the birth of my daughter, Lucy, I started to think about black holes as I was getting into bed. My disability made this rather a slow process, so I had plenty of time. At that date there was no precise definition of which points in space-time lay inside a black hole and which lay outside.

I had already discussed with Roger Penrose the idea of defining a black hole as the set of events from which it was not possible to escape to a large distance. This is now the generally accepted definition. It means that the boundary of the black hole, the event horizon, is formed by rays of light that just fail to get away from the black hole. Instead, they stay forever, hovering on the edge of the black hole. It is like running away from the police and managing to keep one step ahead but not being able to get clear away.

Suddenly I realized that the paths of these light rays could not be approaching one another, because if they were, they must eventually run into each other. It would be like someone else running away from the police in the opposite direction. You would both be caught or, in this case, fall into a black hole. But if these light rays were swallowed up by the black hole, then they could not have been on the boundary of the black hole. So light rays in the event horizon had to be moving parallel to, or away from, each other.

Another way of seeing this is that the event horizon, the boundary of the black hole, is like the edge of a shadow. It is the edge of the light of escape to a great distance, but, equally, it is the edge of the shadow of impending doom. And if you look at the shadow cast by a source at a great distance, such as the sun, you will see that the rays of light on the edge are not approaching each other. If the rays of light that form the event horizon, the boundary of the black hole, can never approach each other, the area of the event horizon could stay the same or increase with time. It could never decrease, because that would mean that at least some of the rays of light in the boundary would have to be approaching each other. In fact, the area would increase whenever matter or radiation fell into the black hole.

Also, suppose two black holes collided and merged together to form a single black hole. Then the area of the event horizon of the final black hole would be greater than the sum of the areas of the event horizons of the original black holes. This nondecreasing property of the event horizon's area placed an important restriction on the possible behavior of black holes. I was so excited with my discovery that I did not get much sleep that night.

The next day I rang up Roger Penrose. He agreed with me. I think, in fact, that he had been aware of this property of the area. However, he had been using a slightly different definition of a black hole. He had not realized that the boundaries of the black hole according to the two definitions would be the same, provided the black hole had settled down to a stationary state.

THE SECOND LAW OF THERMODYNAMICS

The nondecreasing behavior of a black hole's area was very reminiscent of the behavior of a physical quantity called entropy, which measures the degree of disorder of a system. It is a matter of common experience that disorder will tend to increase if things are left to themselves; one has only to leave a house without repairs to see that. One can create order out of disorder—for example, one can paint the house. However, that requires expenditure of energy, and so decreases the amount of ordered energy available.

A precise statement of this idea is known as the second law of thermodynamics. It states that the entropy of an isolated system never decreases with time. Moreover, when two systems are joined together, the entropy of the combined system is greater than the sum of the entropies of the individual systems. For example, consider a system of gas molecules in a box. The molecules can be thought of as little billiard balls continually colliding with each other and bouncing off the walls of the box. Suppose that initially the molecules are all confined to the left-hand side of the box by a partition. If the partition is then removed, the molecules will tend to spread out and occupy both halves of the box. At some later time they could, by chance, all be in the right half or all be back in the left half. However, it is overwhelmingly more probable that there will be roughly equal numbers in the two halves. Such a state is less ordered, or more disordered, than the original state in which all the molecules were in one half. One therefore says that the entropy of the gas has gone up.

Similarly, suppose one starts with two boxes, one containing oxygen molecules and the other containing nitrogen molecules. If one joins the boxes together and removes the intervening wall, the oxygen and the nitrogen molecules will start to mix. At a later time, the most probable state would be to have a thoroughly uniform mixture of oxygen and nitrogen molecules throughout the two boxes. This state would be less ordered, and hence have more entropy, than the initial state of two separate boxes.

The second law of thermodynamics has a rather different status than that of other laws of science. Other laws, such as Newton's law of gravity, for example, are absolute law—that is, they always hold.

On the other hand, the second law is a statistical law—that is, it does not hold always, just in the vast majority of cases. The probability of all the gas molecules in our box being found in one half of the box at a later time is many millions of millions to one, but it could happen.

However, if one has a black hole around, there seems to be a rather easier way of violating the second law: Just throw some matter with a lot of entropy, such as a box of gas, down the black hole. The total entropy of matter outside the black hole would go down. One could, of course, still say that the total entropy, including the entropy inside the black hole, has not gone down. But since there is no way to look inside the black hole, we cannot see how much entropy the matter inside it has. It would be nice, therefore, if there was some feature of the black hole by which observers outside the black hole could tell its entropy; this should increase whenever matter carrying entropy fell into the black hole.

Following my discovery that the area of the event horizon increased whenever matter fell into a black hole, a research student at Princeton named Jacob Bekenstein suggested that the area of the event horizon was a measure of the entropy of the black hole. As matter carrying entropy fell into the black hole, the area of the event horizon would go up, so that the sum of the entropy of matter outside black holes and the area of the horizons would never go down.

This suggestion seemed to prevent the second law of thermodynamics from being violated in most situations. However, there was one fatal flaw: If a black hole has entropy, then it ought also to have a temperature. But a body with a nonzero temperature must emit radiation at a certain rate. It is a matter of common experience that if one heats up a poker in the fire, it glows red hot and emits radiation. However, bodies at lower temperatures emit radiation, too; one just does not normally notice it because the amount is fairly small. This radiation is required in order to prevent violations of the second law. So black holes ought to emit radiation, but by their very definition, black holes are objects that are not supposed to emit anything. It therefore seemed that the area of the event horizon of a black hole could not be regarded as its entropy.

In fact, in 1972 I wrote a paper on this subject with Brandon Carter and an American colleague, Jim Bardeen. We pointed out that, although there were many similarities between entropy and the area of the event horizon, there was this apparently fatal difficulty. I must admit that in writing this paper I was motivated partly by irritation with Bekenstein, because I felt he had misused my discovery of the increase of the area of the event horizon. However, it turned out in the end that he was basically correct, though in a manner he had certainly not expected.

BLACK HOLE RADIATION

In September 1973, while I was visiting Moscow, I discussed black holes with two leading Soviet experts, Yakov Zeldovich and Alexander Starobinsky. They convinced me that, according to the quantum mechanical uncertainty principle, rotating black holes should create and emit particles. I believed their arguments on physical grounds, but I did not like the mathematical way in which they calculated the emission. I therefore set about devising a better mathematical treatment, which I described at an informal seminar in Oxford at the end of November 1973. At that time I had not done the calculations to find out how much would actually be emitted. I was expecting to discover just the radiation that Zeldovich and Starobinsky had predicted from rotating black holes.However, when I did the calculation, I found, to my surprise and annoyance, that even nonrotating black holes should apparently create and emit particles at a steady rate.

At first I thought that this emission indicated that one of the approximations I had used was not valid. I was afraid if Bekenstein found out about it, he would use it as a further argument to support his ideas about the entropy of black holes, which I still did not like. However, the more I thought about it, the more it seemed that the approximations really ought to hold. But what finally convinced me that the emission was real was that the spectrum of the emitted particles was exactly that which would be emitted by a hot body. The black hole was emitting particles at exactly the correct rate to prevent violations of the second law.

Since then, the calculations have been repeated in a number of different forms by other people. They all confirm that a black hole ought to emit particles and radiation as if it were a hot body with a temperature that depends only on the black hole's mass: the higher the mass, the lower the temperature. One can understand this emission in the following way: What we think of as empty space cannot be completely empty because that would mean that all the fields, such as the gravitational field and the electromagnetic field, would have to be exactly zero. However, the value of a field and its rate of change with time are like the position and velocity of a particle. The uncertainty principle implies that the more accurately one knows one of these quantities, the less accurately one can know the other.

So in empty space the field cannot be fixed at exactly zero, because then it would have both a precise value, zero, and a precise rate of change, also zero. Instead, there must be a certain minimum amount of uncertainty, or quantum fluctuations, in the value of a field. One can think of these fluctuations as pairs of particles of light or gravity that appear together at some time, move apart, and then come together again and annihilate each other. These particles are called virtual particles. Unlike real particles, they cannot be observed directly with a particle detector. However, their indirect effects, such as small changes in the energy of electron orbits and atoms, can be measured and agree with the theoretical predictions to a remarkable degree of accuracy.

By conservation of energy, one of the partners in a virtual particle pair will have positive energy and the other partner will have negative energy. The one with negative energy is condemned to be a short-lived virtual particle. This is because real particles always have positive energy in normal situations. It must therefore seek out its partner and annihilate it. However, the gravitational field inside a black hole is so strong that even a real particle can have negative energy there.

It is therefore possible, if a black hole is present, for the virtual particle with negative energy to fall into the black hole and become a real particle. In this case it no longer has to annihilate its partner; its forsaken partner may fall into the black hole as well. But because it has positive energy, it is also possible for it to escape to infinity as a real particle. To an observer at a distance, it will appear to have been emitted from the black hole. The smaller the black hole, the less far the particle with negative energy will have to go before it becomes a real particle. Thus, the rate of emission will be greater, and the apparent temperature of the black hole will be higher.

The positive energy of the outgoing radiation would be balanced by a flow of negative energy particles into the black hole. By Einstein's famous equation $E = mc^2$, energy is equivalent to mass. A flow of negative energy into the black hole therefore reduces its mass. As the black hole loses mass, the area of its event horizon gets smaller, but this decrease in the entropy of the black hole is more than compensated for by the entropy of the emitted radiation, so the second law is never violated.

BLACK HOLE EXPLOSIONS

The lower the mass of the black hole, the higher its temperature is. So as the black hole loses mass, its temperature and rate of emission increase. It therefore loses mass more quickly. What happens when the mass of the black hole eventually becomes extremely small is not quite clear. The most reasonable guess is that it would disappear completely in a tremendous final burst of emission, equivalent to the explosion of millions of H-bombs.

A black hole with a mass a few times that of the sun would have a temperature of only one ten-millionth of a degree above absolute zero. This is much less than the temperature of the microwave radiation that fills the universe, about 2.7 degrees above absolute zero—so such black holes would give off less than they absorb, though even that would be very little. If the universe is destined to go on expand ing forever, the temperature of the microwave radiation will eventually decrease to less than that of such a black hole. The hole will then absorb less than it emits and will begin to lose mass. But, even then, its temperature is so low that it would take about 10^{66} years to evaporate completely. This is much longer than the age of the universe, which is only about 10^{10} years.

On the other hand, as we learned in the last lecture, there might be primordial black holes with a very much smaller mass that were made by the collapse of irregularities in the very early stages of the universe. Such black holes would have a much higher temperature and would be emitting radiation at a much greater rate. A primordial black hole with an initial mass of a thousand million tons would have a lifetime roughly equal to the age of the universe. Primordial black holes with initial masses less than this figure would already have completely evaporated. However, those with slightly greater masses would still be emitting radiation in the form of X rays and gamma rays. These are like waves of light, but with a much shorter wavelength. Such holes hardly deserve the epithet black. They really are white hot, and are emitting energy at the rate of about ten thousand megawatts.

One such black hole could run ten large power stations, if only we could harness its output. This would be rather difficult, however.

The black hole would have the mass of a mountain compressed into the size of the nucleus of an atom. If you had one of these black holes on the surface of the Earth, there would be no way to stop it falling through the floor to the center of the Earth. It would oscillate through the Earth and back, until eventually it settled down at the center. So the only place to put such a black hole, in which one might use the energy that it emitted, would be in orbit around the Earth. And the only way that one could get it to orbit the Earth would be to attract it there by towing a large mass in front of it, rather like a carrot in front of a donkey. This does not sound like a very practical proposition, at least not in the immediate future.

THE SEARCH FOR PRIMORDIAL BLACK HOLES

But even if we cannot harness the emission from these primordial black holes, what are our chances of observing them? We could look for the gamma rays that the primordial black holes emit during most of their lifetime. Although the radiation from most would be very weak because they are far away, the total from all of them might be detectable. We do, indeed, observe such a background of gamma rays. However, this background was probably generated by processes other than primordial black holes. One can say that the observations of the gamma ray background do not provide any positive evidence for primordial black holes. But they tell us that, on average, there cannot be more than three hundred little black holes in every cubic light-year in the universe. This limit means that primordial black holes could make up at most one millionth of the average mass density in the universe.

With primordial black holes being so scarce, it might seem unlikely that there would be one that was near enough for us to observe on its own. But since gravity would draw primordial black holes toward any matter, they should be much more common in galaxies. If they were, say, a million times more common in galaxies, then the nearest black hole to us would probably be at a distance of about a thousand million kilometers, or about as far as Pluto, the farthest known planet. At this distance it would still be very difficult to detect the steady emission of a black hole even if it was ten thousand megawatts.

In order to observe a primordial black hole, one would have to detect several gamma ray quanta coming from the same direction within a reasonable space of time, such as a week.

Otherwise, they might simply be part of the background. But Planck's quantum principle tells us that each gamma ray quantum has a very high energy, because gamma rays have a very high frequency. So to radiate even ten thousand megawatts would not take many quanta. And to observe these few quanta coming from the distance of Pluto would require a larger gamma ray detector than any that have been constructed so far. Moreover, the detector would have to be in space, because gamma rays cannot penetrate the atmosphere.

Of course, if a black hole as close as Pluto were to reach the end of its life and blow up, it would be easy to detect the final burst of emission. But if the black hole has been emitting for the last ten or twenty thousand million years, the chances of it reaching the end of its life within the next few years are really rather small. It might equally well be a few million years in the past or future. So in order to have a reasonable chance of seeing an explosion before your research grant ran out, you would have to find a way to detect any explosions within a distance of about one light-year. You would still have the problem of needing a large gamma ray detector to observe several gamma ray quanta from the explosion. However, in this case, it would not be necessary to determine that all the quanta came from the same direction. It would be enough to observe that they all arrived within a very short time interval to be reasonably confident that they were coming from the same burst.

One gamma ray detector that might be capable of spotting primordial black holes is the entire Earth's atmosphere. (We are, in any case, unlikely to be able to build a larger detector.) When a high-energy gamma ray quantum hits the atoms in our atmosphere, it creates pairs of electrons and positrons. When these hit other atoms, they in turn create more pairs of electrons and positrons. So one gets what is called an electron shower. The result is a form of light called Cerenkov radiation. One can therefore detect gamma ray bursts by looking for flashes of light in the night sky.

Of course, there are a number of other phenomena, such as lightning, which can also give flashes in the sky. However, one could distinguish gamma ray bursts from such effects by observing flashes simultaneously at two or more thoroughly widely separated locations. A search like this has been carried out by two scientists from Dublin, Neil Porter and Trevor Weekes, using telescopes in Arizona. They found a number of flashes but none that could be definitely ascribed to gamma ray bursts from primordial black holes.

Even if the search for primordial black holes proves negative, as it seems it may, it will still give us important information about the very early stages of the universe. If the early universe had been chaotic or irregular, or if the pressure of matter had been low, one would have expected it to produce many more primordial black holes than the limit set by our observations of the gamma ray background. It is only if the early universe was very smooth and uniform, and with a high pressure, that one can explain the absence of observable numbers of primordial black holes.

GENERAL RELATIVITY AND QUANTUM MECHANICS

Radiation from black holes was the first example of a prediction that depended on both of the great theories of this century, general relativity and quantum mechanics. It aroused a lot of opposition initially because it upset the existing viewpoint: "How can a black hole emit anything?" When I first announced the results of my calculations at a conference at the Rutherford Laboratory near Oxford, I was greeted with general incredulity. At the end of my talk the chairman of the session, John G. Taylor from Kings College, London, claimed it was all nonsense. He even wrote a paper to that effect.

However, in the end most people, including John Taylor, have come to the conclusion that black holes must radiate like hot bodies if our other ideas about general relativity and quantum mechanics are correct. Thus even though we have not yet managed to find a primordial black hole, there is fairly general agreement that if we did, it would have to be emitting a lot of gamma and X rays. If we do find one, I will get the Nobel Prize.

The existence of radiation from black holes seems to imply that gravitational collapse is not as final and irreversible as we once thought. If an astronaut falls into a black hole, its mass will increase. Eventually, the energy equivalent of that extra mass will be returned to the universe in the form of radiation. Thus, in a sense, the astronaut will be recycled. It would be a poor sort of immortality, however, because any personal concept of time for the astronaut would almost certainly come to an end as he was crushed out of existence inside the black hole. Even the types of particle that were eventually emitted by the black hole would in general be different from those that made up the astronaut. The only feature of the astronaut that would survive would be his mass or energy.

The approximations I used to derive the emission from black holes should work well when the black hole has a mass greater than a fraction of a gram. However, they will break down at the end of the black hole's life, when its mass gets very small. The most likely outcome seems to be that the black hole would just disappear, at least from our region of the universe. It would take with it the astronaut and any singularity there might be inside the black hole. This was the first indication that quantum mechanics might remove the singularities that were predicted by classical general relativity. However, the methods that I and other people were using in 1974 to study the quantum effects of gravity were not able to answer questions such as whether singularities would occur in quantum gravity.

From 1975 onward, I therefore started to develop a more powerful approach to quantum gravity based on Feynman's idea of a sum over histories. The answers that this approach suggests for the origin and fate of the universe will be described in the next two lectures. We shall see that quantum mechanics allows the universe to have a beginning that is not a singularity. This means that the laws of physics need not break down at the origin of the universe. The state of the universe and its contents, like ourselves, are completely determined by the laws of physics, up to the limit set by the uncertainty principle. So much for free will.

LECTURE 5
THE ORIGIN AND FATE OF THE UNIVERSE

Throughout the 1970s I had been working mainly on black holes. However, in 1981 my interest in questions about the origin of the universe was reawakened when I attended a conference on cosmology in the Vatican. The Catholic church had made a bad mistake with Galileo when it tried to lay down the law on a question of science, declaring that the sun went around the Earth. Now, centuries later, it had decided it would be better to invite a number of experts to advise it on cosmology.

At the end of the conference the participants were granted an audience with the pope. He told us that it was okay to study the evolution of the universe after the big bang, but we should not inquire into the big bang itself because that was the moment of creation and therefore the work of God.

I was glad then that he did not know the subject of the talk I had just given at the conference. I had no desire to share the fate of Galileo; I have a lot of sympathy with Galileo, partly because I was born exactly three hundred years after his death.

THE HOT BIG BANG MODEL

In order to explain what my paper was about, I shall first describe the generally accepted history of the universe, according to what is known as the "hot big bang model." This assumes that the universe is described by a Friedmann model, right back to the big bang. In such models one finds that as the universe expands, the temperature of the matter and radiation in it will go down. Since temperature is simply a measure of the average energy of the particles, this cooling of the universe will have a major effect on the matter in it. At very high temperatures, particles will be moving around so fast that they can escape any attraction toward each other caused by the nuclear or electromagnetic forces. But as they cooled off, one would expect particles that attract each other to start to clump together.

At the big bang itself, the universe had zero size and so must have been infinitely hot. But as the universe expanded, the temperature of the radiation would have decreased. One second after the big bang it would have fallen to about ten thousand million degrees. This is about a thousand times the temperature at the center of the sun, but temperatures as high as this are reached in H-bomb explosions. At this time the universe would have contained mostly photons, electrons, and neutrinos and their antiparticles, together with some protons and neutrons.

As the universe continued to expand and the temperature to drop, the rate at which electrons and the electron pairs were being produced in collisions would have fallen below the rate at which they were being destroyed by annihilation. So most of the electrons and antielectrons would have annihi lated each other to produce more photons, leaving behind only a few electrons.

About one hundred seconds after the big bang, the temperature would have fallen to one thousand million degrees, the temperature inside the hottest stars. At this temperature, protons and neutrons would no longer have sufficient energy to escape the attraction of the strong nuclear force. They would start to combine together to produce the nuclei of atoms of deuterium, or heavy hydrogen, which contain one proton and one neutron. The deuterium nuclei would then have combined with more protons and neutrons to make helium nuclei, which contained two protons and two neutrons. There would also be small amounts of a couple of heavier elements, lithium and beryllium.

One can calculate that in the hot big bang model about a quarter of the protons and neutrons would have been converted into helium nuclei, along with a small amount of heavy hydrogen and other elements. The remaining neutrons would have decayed into protons, which are the nuclei of ordinary hydrogen atoms. These predictions agree very well with what is observed.

The hot big bang model also predicts that we should be able to observe the radiation left over from the hot early stages. However, the temperature would have been reduced to a few degrees above absolute zero by the expansion of the universe. This is the explanation of the microwave background of radiation that was discovered by Penzias and Wilson in 1965. We are therefore thoroughly confident that we have the right picture, at least back to about one second after the big bang. Within only a few hours of the big bang, the production of helium and other elements would have stopped. And after that, for the next million years or so, the universe would have just continued expanding, without anything much happening. Eventually, once the temperature had dropped to a few thousand degrees, the electrons and nuclei would no longer have had enough energy to overcome the electromagnetic attraction between them. They would then have started combining to form atoms.

The universe as a whole would have continued expanding and cooling. However, in regions that were slightly denser than average, the expansion would have been slowed down by extra gravitational attraction. This would eventually stop expansion in some regions and cause them to start to recollapse. As they were collapsing, the gravitational pull of matter outside these regions might start them rotating slightly. As the collapsing region got smaller, it would spin faster—just as skaters spinning on ice spin faster as the draw in their arms. Eventually, when the region got small enough, it would be spinning fast enough to balance the attraction of gravity. In this way, disklike rotating galaxies were born.

As time went on, the gas in the galaxies would break up into smaller clouds that would collapse under their own gravity. As these contracted, the temperature of the gas would increase until it became hot enough to start nuclear reactions. These would convert the hydrogen into more helium, and the heat given off would raise the pressure, and so stop the clouds from contracting any further. They would remain in this state for a long time as stars like our sun, burning hydrogen into helium and radiating the energy as heat and light.

More massive stars would need to be hotter to balance their stronger gravitational attraction. This would make the nuclear fusion reactions proceed so much more rapidly that they would use up their hydrogen in as little as a hundred million years. They would then contract slightly and, as they heated up further, would start to convert helium into heavier elements like carbon or oxygen. This, however, would not release much more energy, so a crisis would occur, as I described in my lecture on black holes.

What happens next is not completely clear, but it seems likely that the central regions of the star would collapse to a very dense state, such as a neutron star or black hole. The outer regions of the star may get blown off in a tremendous explosion called a supernova, which would outshine all the other stars in the galaxy. Some of the heavier elements produced near the end of the star's life would be flung back into the gas in the galaxy. They would provide some of the raw material for the next generation of stars.

Our own sun contains about 2 percent of these heavier elements because it is a second–or third–generation star. It was formed some five thousand million years ago out of a cloud of rotating gas containing the debris of earlier supernovas. Most of the gas in that cloud went to form the sun or got blown away. However, a small amount of the heavier elements collected together to form the bodies that now orbit the sun as planets like the Earth.

OPEN QUESTIONS

This picture of a universe that started off very hot and cooled as it expanded is in agreement with all the observational evidence that we have today. Nevertheless, it leaves a number of important questions unanswered. First, why was the early universe so hot? Second, why is the universe so uniform on a large scale—why does it look the same at all points of space and in all directions?

Third, why did the universe start out with so nearly the critical rate of expansion to just avoid recollapse? If the rate of expansion one second after the big bang had been smaller by even one part in a hundred thousand million million, the universe would have recollapsed before it ever reached its present size. On the other hand, if the expansion rate at one second had been larger by the same amount, the universe would have expanded so much that it would be effectively empty now.

Fourth, despite the fact that the universe is so uniform and homogenous on a large scale, it contains local lumps such as stars and galaxies. These are thought to have developed from small differences in the density of the early universe from one region to another. What was the origin of these density fluctuations?

The general theory of relativity, on its own, cannot explain these features or answer these questions. This is because it predicts that the universe started off with infinite density at the big bang singularity. At the singularity, general relativity and all other physical laws would break down. One cannot predict what would come out of the singularity. As I explained before, this means that one might as well cut any events before the big bang out of the theory, because they can have no effect on what we observe.

Space-time would have a boundary—a beginning at the big bang. Why should the universe have started off at the big bang in just such a way as to lead to the state we observe today? Why is the universe so uniform, and ex panding at just the critical rate to avoid recollapse? One would feel happier about this if one could show that quite a number of different initial configurations for the universe would have evolved to produce a universe like the one we observe. If this is the case, a universe that developed from some sort of random initial conditions should contain a number of regions that are like what we observe. There might also be regions that were very different. However, these regions would probably not be suitable for the formation of galaxies and stars. These are essential prerequisites for the development of intelligent life, at least as we know it. Thus, these regions would not contain any beings to observe that they were different.

When one considers cosmology, one has to take into account the selection principle that we live in a region of the universe that is suitable for intelligent life. This fairly obvious and elementary consideration is sometimes called the anthropic principle. Suppose, on the other hand, that the initial state of the universe had to be chosen extremely carefully to lead to something like what we see around us. Then the universe would be unlikely to contain any region in which life would appear.

In the hot big bang model that I described earlier, there was not enough time in the early universe for heat to have flowed from one region to another. This means that different regions of the universe would have had to have started out with exactly the same temperature in order to account for the fact that the microwave background has the same temperature in every direction we look. Also, the initial rate of expansion would have had to be chosen very precisely for the universe not to have recollapsed before now. This means that the initial state of the universe must have been very carefully chosen indeed if the hot big bang model was correct right back to the beginning of time. It would be very difficult to explain why the universe should have begun in just this way, except as the act of a God who intended to create beings like us.

THE INFLATIONARY MODEL

In order to avoid this difficulty with the very early stages of the hot big bang model, Alan Guth at the Massachusetts Institute of Technology put forward a new model. In this, many different initial configurations could have evolved to something like the present universe. He suggested that the early universe might have had a period of very rapid, or exponential, expansion. This expansion is said to be inflationary—an analogy with the inflation in prices that occurs to a greater or lesser degree in every country. The world record for price inflation was probably in Germany after the first war, when the price of a loaf of bread went from under a mark to millions of marks in a few months. But the inflation we think may have occurred in the size of the universe was much greater even than that—a million million million million million times in only a tiny fraction of a second. Of course, that was before the present government.

Guth suggested that the universe started out from the big bang very hot. One would expect that at such high temperatures, the strong and weak nuclear forces and the electromagnetic force would all be unified into a single force. As the universe expanded, it would cool, and particle energies would go down. Eventually there would be what is called a phase transition, and the symmetry between the forces would be broken. The strong force would become different from the weak and electromagnetic forces. One common example of a phase transition is the freezing of water when you cool it down. Liquid water is symmetrical, the same at every point and in every direction. However, when ice crystals form, they will have definite positions and will be lined up in some direction. This breaks the symmetry of the water.

In the case of water, if one is careful, one can "supercool" it. That is, one can reduce the temperature below the freezing point—0 degrees centigrade—without ice forming. Guth suggested that the universe might behave in a similar way: The temperature might drop below the critical value without the symmetry between the forces being broken. If this happened, the universe would be in an unstable state, with more energy than if the symmetry had been broken. This special extra energy can be shown to have an antigravitational effect. It would act just like a cosmological constant.

Einstein introduced the cosmological constant into general relativity when he was trying to construct a static model of the universe. However,in this case, the universe would already be expanding. The repulsive effect of this cosmological constant would therefore have made the universe expand at an everincreasing rate. Even in regions where there were more matter particles than average, the gravitational attraction of the matter would have been outweighed by the repulsion of the effective cosmological constant. Thus, these regions would also expand in an accelerating inflationary manner.

As the universe expanded, the matter particles got farther apart. One would be left with an expanding universe that contained hardly any particles. It would still be in the supercooled state, in which the symmetry between the forces is not broken. Any irregularities in the universe would simply have been smoothed out by the expansion, as the wrinkles in a balloon are smoothed away when you blow it up. Thus, the present smooth and uniform state of the universe could have evolved from many different nonuniform initial states. The rate of expansion would also tend toward just the critical rate needed to avoid recollapse.

Moreover, the idea of inflation could also explain why there is so much matter in the universe. There are something like 1,080 particles in the region of the universe that we can observe. Where did they all come from? The answer is that, in quantum theory, particles can be created out of energy in the form of particle/antiparticle pairs. But that just raises the question of where the energy came from. The answer is that the total energy of the universe is exactly zero.

The matter in the universe is made out of positive energy. However, the matter is all attracting itself by gravity. Two pieces of matter that are close to each other have less energy than the same two pieces a long way apart. This is because you have to expend energy to separate them. You have to pull against the gravitational force attracting them together. Thus, in a sense, the gravitational field has negative energy. In the case of the whole universe, one can show that this negative gravitational energy exactly cancels the positive energy of the matter. So the total energy of the universe is zero.

Now, twice zero is also zero. Thus, the universe can double the amount of positive matter energy and also double the negative gravitational energy without violation of the conservation of energy. This does not happen in the normal expansion of the universe in which the matter energy density goes down as the universe gets bigger. It does happen, however, in the inflationary expansion, because the energy density of the supercooled state remains constant while the universe expands. When the universe doubles in size, the positive matter energy and the negative gravitational energy both double, so the total energy remains zero. During the inflationary phase, the universe increases its size by a very large amount. Thus, the total amount of energy available to make particles becomes very large. As Guth has remarked, "It is said that there is no such thing as a free lunch. But the universe is the ultimate free lunch."

THE END OF INFLATION

The universe is not expanding in an inflationary way today. Thus, there had to be some mechanism that would eliminate the very large effective cosmological constant. This would change the rate of expansion from an accelerated one to one that is slowed down by gravity, as we have today. As the universe expanded and cooled, one might expect that eventually the symmetry between the forces would be broken, just as supercooled water always freezes in the end. The extra energy of the unbroken symmetry state would then be released and would reheat the universe. The universe would then go on to expand and cool, just like the hot big bang model. However, there would now be an explanation of why the universe was expanding at exactly the critical rate and why different regions had the same temperature.

In Guth's original proposal, the transition to broken symmetry was supposed to occur suddenly, rather like the appearance of ice crystals in very cold water. The idea was that "bubbles" of the new phase of broken symmetry would have formed in the old phase, like bubbles of steam surrounded by boiling water. The bubbles were supposed to expand and meet up with each other until the whole universe was in the new phase. The trouble was, as I and several other people pointed out, the universe was expanding so fast that the bubbles would be moving away from each other too rapidly to join up. The universe would be left in a very nonuniform state, with some regions having symmetry between the different forces. Such a model of the universe would not correspond to what we see.

In October 1981 I went to Moscow for a conference on quantum gravity. After the conference, I gave a seminar on the inflationary model and its problems at the Sternberg Astronomical Institute. In the audience was a young Russian, Andrei Linde. He said that the difficulty with the bubbles not joining up could be avoided if the bubbles were very big. In this case, our region of the universe could be contained inside a single bubble. In order for this to work, the change from symmetry to broken symmetry must have taken place very slowly inside the bubble, but this is quite possible according to grand unified theories.

Linde's idea of a slow breaking of symmetry was very good, but I pointed out that his bubbles would have been bigger than the size of the universe at the time. I showed that instead the symmetry would have broken everywhere at the same time, rather than just in-side bubbles. This would lead to a uniform universe, like we observe. The slow symmetry breaking model was a good attempt to explain why the universe is the way it is. However, I and several other people showed that it predicted much greater variations in the microwave background radiation than are observed. Also, later work cast doubt on whether there would have been the right kind of phase transition in the early universe. A better model, called the chaotic inflationary model, was introduced by Linde in 1983. This doesn't de pend on phase transitions, and it can give us the right size of varia-tions of the microwave background. The inflationary model showed that the present state of the universe could have arisen from quite a large number of different initial configurations. It cannot be the case, however, that every initial configuration would have led to a universe like the one we observe. So even the inflationary model does not tell us why the initial configuration was such as to produce what we observe. Must we turn to the anthropic principle for an explanation? Was it all just a lucky chance? That would seem a counsel of despair, a negation of all our hopes of understanding the underlying order of the universe.

QUANTUM GRAVITY

In order to predict how the universe should have started off, one needs laws that hold at the beginning of time. If the classical theory of general relativity was correct, the singularity theorem showed that the beginning of time would have been a point of infinite density and curvature. All the known laws of science would break down at such a point. One might suppose that there were new laws that held at singularities, but it would be very difficult even to formulate laws at such badly behaved points and we would have no guide from observations as to what those laws might be. However, what the singularity theorems really indicate is that the gravitational field becomes so strong that quantum gravitational effects become important: Classical theory is no longer a good description of the universe. So one has to use a quantum theory of gravity to discuss the very early stages of the universe. As we shall see, it is possible in the quantum theory for the ordinary laws of science to hold everywhere, including at the beginning of time. It is not necessary to postulate new laws for singularities, because there need not be any singularities in the quantum theory.

We don't yet have a complete and consistent theory that combines quantum mechanics and gravity. However, we are thoroughly certain of some features that such a unified theory should have. One is that it should incorporate Feynman's proposal to formulate quantum theory in terms of a sum over histories. In this approach, a particle going from A to B does not have just a single history as it would in a classical theory. Instead, it is supposed to follow every possible path in space-time. With each of these histories, there are associated a couple of numbers, one representing the size of a wave and the other representing its position in the cycle—its phase.

The probability that the particle, say, passes through some particular point is found by adding up the waves associated with every possible history that passes through that point. When one actually tries to perform these sums, however, one runs into severe technical problems. The only way around these is the following peculiar pre scription: One must add up the waves for particle histories that are not in the real time that you and I experience but take place in imaginary time.

Imaginary time may sound like science fiction, but it is in fact a well-defined mathematical concept. To avoid the technical difficulties with Feynman's sum over histories, one must use imaginary time. This has an interesting effect on space-time: The distinction between time and space disappears completely. A space-time in which events have imaginary values of the time coordinate is said to be Euclidean because the metric is positive definite.

In Euclidean space-time there is no difference between the time direction and directions in space. On the other hand, in real space-time, in which events are labeled by real values of the time coordinate, it is easy to tell the difference. The time direction lies within the light cone, and space directions lie outside. One can regard the use of imaginary time as merely a mathematical device—or trick—to calculate answers about real space-time. However, there may be more to it than that. It may be that Euclidean space-time is the fundamental concept and what we think of as real space-time is just a figment of our imagination.

When we apply Feynman's sum over histories to the universe, the analogue of the history of a particle is now a complete curved space-time which represents the history of the whole universe. For the technical reasons mentioned above, these curved space-times must be taken to be Euclidean. That is, time is imaginary and is indistinguishable from directions in space. To calculate the probability of finding a real space-time with some certain property, one adds up the waves associated with all the histories in imaginary time that have that property. One can then work out what the probable history of the universe would be in rel time.

THE NO BOUNDARY CONDITION

In the classical theory of gravity, which is based on real space-time, there are only two possible ways the universe can behave. Either it has existed for an infinite time, or else it had a beginning at a singularity at some finite time in the past. In fact, the singu larity theorems show it must be the second possibility. In the quantum theory of gravity, on the other hand, a third possibility arises. Because one is using Euclidean space-times, in which the time direction is on the same footing as directions in space, it is possible for space-time to be finite in extent and yet to have no singularities that formed a boundary or edge. Space-time would be like the surface of the Earth, only with two more dimensions. The surface of the Earth is finite in extent but it doesn't have a boundary or edge. If you sail off into the sunset, you don't fall off the edge or run into a singularity. I know, because I have been around the world.

If Euclidean space-times direct back to infinite imaginary time or else started at a singularity, we would have the same problem as in the classical theory of specifying the initial state of the universe. God may know how the universe began, but we cannot give any particular reason for thinking it began one way rather than another. On the other hand, the quantum theory of gravity has opened up a new possibility. In this, there would be no boundary to space-time. Thus, there would be no need to specify the behavior at the boundary. There would be no singularities at which the laws of science broke down and no edge of space-time at which one would have to appeal to God or some new law to set the boundary conditions for space-time. One could say: "The boundary condition of the universe is that it has no boundary." The universe would be completely self-contained and not affected by anything outside itself. It would be neither created nor destroyed. It would just be.

It was at the conference in the Vatican that I first put forward the suggestion that maybe time and space together formed a surface that was finite in size but did not have any boundary or edge. My paper was rather mathematical, however, so its implications for the role of God in the creation of the universe were not noticed at the time-just as well for me. At the time of the Vatican conference, I did not know how to use a no boundary idea to make predictions about the universe. However, I spent the following summer at the University of California, Santa Barbara. There, a friend and colleague of mine, Jim Hartle, worked out with me what conditions the universe must satisfy if space-time had no boundary.

I should emphasize that this idea that time and space should be finite without boundary is just a proposal. It cannot be deduced from some other principle. Like any other scientific theory, it may initially be put forward for aesthetic or metaphysical reasons, but the real test is whether it makes predictions that agree with observation. This, however, is difficult to determine in the case of quantum gravity, for two reasons. First, we are not yet sure exactly which theory successfully combines general relativity and quantum mechanics, though we know quite a lot about the form such a theory must have. Second, any model that described the whole universe in detail would be much too complicated mathematically for us to be able to calculate exact predictions. One therefore has to make approximations—and even then, the problem of extracting predictions remains a difficult one.

One finds, under the no boundary proposal, that the chance of the universe being found to be following most of the possible histories is negligible. But there is a particular family of histories that are much more probable than the others. These histories may be pictured as being like the surface of the Earth, with a distance from the North Pole representing imaginary time; the size of a circle of latitude would represent the spatial size of the universe. The universe starts at the North Pole as a single point. As one moves south, the circle of latitude get bigger, corresponding to the universe expanding with imaginary time. The universe would reach a maximum size at the equator and would contract again to a single point at the South Pole. Even though the universe would have zero size at the North and South poles, these points would not be singularities any more than the North and South poles on the Earth are singular. The laws of science will hold at the beginning of the universe, just as they do at the North and South poles on the Earth.

The history of the universe in real time, however, would look very different. It would appear to start at some minimum size, equal to the maximum size of the history in imaginary time. The universe would then expand in real time like the inflationary model. However, one would not now have to assume that the universe was created somehow in the right sort of state. The universe would expand to a very large size, but eventually it would collapse again into what looks like a singularity in real time. Thus, in a sense, we are still all doomed, even if we keep away from black holes. Only if we could picture the universe in terms of imaginary time would there be no singularities.

The singularity theorems of classical general relativity showed that the universe must have a beginning, and that this beginning must be described in terms of quantum theory. This in turn led to the idea that the universe could be finite in imaginary time, but without boundaries or singularities. When one goes back to the real time in which we live, however, there will still appear to be singularities. The poor astronaut who falls into a black hole will still come to a sticky end. It is only if he could live in imaginary time that he would encounter no singularities.

This might suggest that the so-called imaginary time is really the fundamental time, and that what we call real time is something we create just in our minds. In real time, the universe has a beginning and an end at singularities that form a boundary to space-time and at which the laws of science break down. But in imaginary time, there are no singularities or boundaries. So maybe what we call imaginary time is really more basic, and what we call real time is just an idea that we invent to help us describe what we think the universe is like. But according to the approach I described in the first lecture, a scientific theory is just a mathematical model we make to describe our observations. It exists only in our minds. So it does not have any meaning to ask: Which is real, "real" or "imaginary" time? It is simply a matter of which is a more useful description.

The no boundary proposal seems to predict that, in real time, the universe should behave like the inflationary models. A particularly interesting problem is the size of the small departures from uniform density in the early universe. These are thought to have led to the formation first of the galaxies, then of stars, and finally of beings like us. The uncertainty principle implies that the early universe cannot have been completely uniform. Instead, there must have been some uncertainties or fluctuations in the positions and velocities of the particles. Using the no boundary condition, one finds that the universe must have started off with just the minimum possible nonuniformity allowed by the uncertainty principle.

The universe would have then undergone a period of rapid expansion, like in the inflationary models. During this period, the initial nonuniformities would have been amplified until they could have been big enough to explain the origin of galaxies. Thus, all the complicated structures that we see in the universe might be explained by the no boundary condition for the universe and the uncertainty principle of quantum mechanics.

The idea that space and time may form a closed surface without boundary also has profound implications for the role of God in the affairs of the universe. With the success of scientific theories in describing events, most people have come to believe that God allows the universe to evolve according to a set of laws. He does not seem to intervene in the universe to break these laws. However, the laws do not tell us what the universe should have looked like when it started. It would still be up to God to wind up the clockwork and choose how to start it off. So long as the universe had a beginning that was a singularity, one could suppose that it was created by an outside agency. But if the universe is really completely self-contained, having no boundary or edge, it would be neither created nor destroyed. It would simply be. What place, then, for a creator?

LECTURE *6*
THE DIRECTION OF TIME

In his book, The *Go Between*, L. P. Hartley wrote, "The past is a foreign country. They do things differently there-but why is the past so different from the future? Why do we remember the past, but not the future?" In other words, why does time go forward? Is this connected with the fact that the universe is expanding?

C, P, T

The laws of physics do not distinguish between the past and the future. More precisely, the laws of physics are unchanged under the combination of operations known as C, P, and T. (C means changing particles for antiparticles. P means taking the mirror image so left and right are swapped for each other. And T means reversing the direction of motion of all particles—in effect, running the motion backward.) The laws of physics that govern the behavior of matter under all normal situations are unchanged under the operations C and P on their own. In other words, life would be just the same for the inhabitants of another planet who were our mirror images and who were made of antimatter. If you meet someone from another planet and he holds out his left hand, don't shake it. He might be made of antimatter. You would both disappear in a tremendous flash of light. If the laws of physics are unchanged by the combination of operations C and P, and also by the combination C, P, and T, they must also be unchanged under the operation T alone. Yet, there is a big difference between the forward and backward directions of time in ordinary life. Imagine a cup of water falling off a table and breaking in pieces on the floor. If you take a film of this, you can easily tell whether it is being run forward or backward. If you run it backward, you will see the pieces suddenly gather themselves together off the floor and jump back to form a whole cup on the table. You can tell that the film is being run backward because this kind of behavior is never observed in ordinary life. If it were, the crockery manufacturers would go out of business.

THE ARROWS OF TIME

The explanation that is usually given as to why we don't see broken cups jumping back onto the table is that it is forbidden by the second law of thermodynamics. This says that disorder or entropy always increases with time. In other words, it is Murphy's Law—things get worse. An intact cup on the table is a state of high order, but a broken cup on the floor is a disordered state. One can therefore go from the whole cup on the table in the past to the broken cup on the floor in the future, but not the other way around.

The increase of disorder or entropy with time is one example of what is called *an arrow of time*, something that gives a direction to time and distinguishes the past from the future. There are at least three different arrows of time. First, there is the thermodynamic arrow of time—the direction of time in which disorder or entropy increases. Second, there is the psychological arrow of time. This is the direction in which we feel time passes—the direction of time in which we remember the past, but not the future. Third, there is the cosmological arrow of time. This is the direction of time in which the universe is expanding rather than contracting.

I shall argue the the pyschological arrow is determined by the thermodynamic arrow and that these two arrows always point in the same direction. If one makes the no boundary assumption for the universe, they are related to the cosmological arrow of time, though they may not point in the same direction. However, I shall argue that it is only when they agree with the cosmological arrow that there will be intelligent beings who can ask the question: Why does disorder increase in the same direction of time as that in which the universe expands?

THE THERMODYNAMIC ARROW

I shall talk first about the thermodynamic arrow of time. The second law of thermodynamics is based on the fact that there are many more disordered states than there are ordered ones. For example, consider the pieces of a jigsaw in a box. There is one, and only one, arrangement in which the pieces make a complete picture. On the other hand, there are a very large number of arrangements in which the pieces are disordered and don't make a picture.

Suppose a systems starts out in one of the small number of ordered states. As time goes by, the system will evolve according to the laws of physics and its state will change. At a later time, there is a high probability that it will be in a more disordered state, simply because there are so many more disordered states. Thus, disorder will tend to increase with time if the system obeys an initial condition of high order.

Suppose the pieces of the jigsaw start off in the ordered arrangement in which they form a picture. If you shake the box, the pieces will take up another arrangement. This will probably be a disordered arrangement in which the pieces don't form a proper picture, simply because there are so many more disordered arrangements. Some groups of pieces may still form parts of the picture, but the more you shake the box, the more likely it is that these groups will get broken up. The pieces will take up a completely jumbled state in which they don't form any sort of picture. Thus, the disorder of the pieces will probably increase with time if they obey the initial condition that they start in a state of high order.

Suppose, however, that God decided that the universe should finish up at late times in a state of high order but it didn't matter what state it started in. Then, at early times the universe would probably be in a disordered state, and disorder would decrease with time. You would have broken cups gathering themselves together and jumping back on the table. However, any human beings who observing the cups would be living in a universe in which disorder decreased with time. I shall argue that such beings would have a psychological arrow of time that was backward. That is, they would remember thence at late times and not remember thence at early times.

THE PSYCHOLOGICAL ARROW

It is rather difficult to talk about human memory because we don't know how the brain works in detail. We do, however, know all about how computer memories work. I shall therefore discuss the psychological arrow of time for computers. I think it is reasonable to assume that the arrow for computers is the same as that for human. If it were not, one could make a killing on the stock exchange by having a computer that would remember tomorrow's prices.

A computer memory is basically some device that can be in either one of two states. An example would be a superconducting loop of wire. If there is an electric current flowing in the loop, it will continue to flow because there is no resistance. On the other hand, if there is no current, the loop will continue without a current. One can label the two states of the memory "one" and "zero."

Before an item is recorded in the memory, the memory is in a disordered state with equal probabilities for one and zero. After the memory interacts with the system to be remembered, it will definitely be in one state or the other, according to the state of the system. Thus, the memory passes from a disordered state to an ordered one. However, in order to make sure that the memory is in the right state, it is necessary to use a certain amount of energy. This energy is dissipated as heat and increases the amount of disorder in the universe. One can show that this increase of disorder is greater than the increase in the order of the memory. Thus, when a computer records an item in memory, the total amount of disorder in the universe goes up.

The direction of time in which a computer remembers the past is the same as that in which disorder increases. This means that our subjective sense of the direction of time, the psychological arrow of time, is determined by the thermodynamic arrow of time. This makes the second law of thermodynamics almost trivial. Disorder increases with time because we measure time in the direction in which disorder increases. You can't have a safer bet than that.

THE BOUNDARY CONDITIONS OF THE UNIVERSE

But why should the universe be in a state of high order at one end of time, the end that we call the past? Why was it not in a state of complete disorder at all times? After all, this might seem more probable. And why is the direction of time in which disorder increases the same as that in which the universe expands? One possible answer is that God simply chose that the universe should be in a smooth and ordered state at the beginning of the expansion phase. We should not try to understand why or question His reasons because the beginning of the universe was the work of God. But the whole history of the universe can be said to be the work of God.

It appears that the universe evolves according to well-defined laws. These laws may or may not be ordained by God, but it seems that we can discover and understand them. Is it, therefore, unreasonable to hope that the same or similar laws may also hold at the beginning of the universe? In the classical theory of general relativity, the beginning of the universe has to be a singularity of infinite density in space-time curvature. Under such conditions, all the known laws of physics would break down. Thus, one could not use them to predict how the universe would begin.

The universe could have started out in a very smooth and ordered state. This would have led to well-defined thermodynamic and cosmological arrows of time, like we observe. But it could equally well have started out in a very lumpy and disordered state. In this case, the universe would already be in a state of complete disorder, so disorder could not increase with time. It would either stay constant, in which case there would be no well-defined thermodynamic arrow of time, or it would decrease, in which case the thermodynamic arrow of time would point in the opposite direction to the cosmological arrow. Neither of these possibilities would agree with what we observe.

As I mentioned, the classical theory of general relativity predicts that the universe should begin with a singularity where the curvature of space-time is infinite. In fact, this means that classical general relativity predicts its own downfall. When the curvature of space-time becomes large, quantum gravitationa effects will become important and the classical theory will cease to be a good description of the universe. One has to use the quantum theory of gravity to understand how the universe began.

In a quantum theory of gravity, one considers all possible histories of the universe. Associated with each history, there are a couple of numbers. One represents the size of a wave and the other the face of the wave, that is, whether the wave is at a crest or a trough. The probability of the universe having a particular property is given by adding up the waves for all the histories with that property. The histories would be curved spaces that would represent the evolution of the universe in time. One would still have to say how the possible histories of the universe would behave at the boundary of space-time in the past. We do not and cannot know the boundary conditions of the universe in the past. However, one could avoid this difficulty if the boundary condition of the universe is that it has no boundary. In other words, all the possible histories are finite in extent but have no boundaries, edges, or singularities. They are like the surface of the Earth, but with two more dimensions. In that case, the beginning of time would be a regular smooth point of space-time. This means that the universe would have begun its expansion in a very smooth and ordered state. It could not have been completely uniform because that would violate the uncertainty principle of quantum theory. There had to be small fluctuations in the density and velocities of particles. The no boundary condition, however, would imply that these fluctuations were as small as they could be, consistent with the uncertainty principle.

The universe would have started off with a period of exponential or "inflationary" expansion. In this, it would have increased its size by a very large factor. During this expansion, the density fluctuations would have remained small at first, but later would have started to grow. Regions in which the density was slightly higher than average would have had their expansion slowed down by the gravitational attraction of the extra mass. Eventually, such regions would stop expanding, and would collapse to form galaxies, stars, and beings like us.

The universe would have started in a smooth and ordered state and would become lumpy and disordered as time went on. This would explain the existence of the thermodynamic arrow of time.

The universe would start in a state of high order and would become more disordered with time. As I showed earlier, the psychological arrow of time points in the same direction as the thermodynamic arrow. Our subjective sense of time would therefore be that in which the universe is expanding, rather than the opposite direction, in which it would be contracting.

DOES THE ARROW OF TIME REVERSE?

But what would happen if and when the universe stopped expanding and began to contract again? Would the thermodynamic arrow reverse and disorder begin to decrease with time? This would lead to all sorts of science-fiction-like possibilities for people who survived from the expanding to the contracting phase. Would they see broken cups gathering themselves together off the floor and jumping back on the table? Would they be able to remember tomorrow's prices and make a fortune on the stock market?

It might seem a bit academic to worry about what would happen when the universe collapses again, as it will not start to contract for at least another ten thousand million years. But there is a quicker way to find out what will happen: Jump into a black hole. The collapse of a star to form a black hole is rather like the later stages of the collapse of the whole universe. Thus, if disorder were to decrease in the contracting phase of the universe, one might also expect it to decrease inside a black hole. So perhaps an astronaut who fell into a black hole would be able to make money at roulette by remembering where the ball went before he placed his bet. Unfortunately, however, he would not have long to play before he was turned to spaghetti by the very strong gravitational fields. Nor would he be able to let us know about the reversal of the thermodynamic arrow, or even bank his winnings, because he would be trapped behind the event horizon of the black hole.

At first, I believed that disorder would decrease when the universe recollapsed. This was because I thought that the universe had to return to a smooth and ordered state when it became small again. This would have meant that the contracting phase was like the time reverse of the expanding phase. People in the contracting phase would live their lives backward. They would die before they were born and would get younger as the universe contracted. This idea is attractive because it would mean a nice symmetry between the expanding and contracting phases. However, one cannot adopt it on its own, independent of other ideas about the universe. The question is: Is it implied by the no boundary condition or is it inconsistent with that condition?

As I mentioned, I thought at first that the no boundary condition did indeed imply that disorder would decrease in the contracting phase. This was based on work on a simple model of the universe in which the collapsing phase looked like the time reverse of the expanding phase. However, a colleague of mine, Don Page, pointed out that the no boundary condition did not require the contracting phase necessarily to be the time reverse of the expanding phase. Further, one of my students, Raymond Laflamme, found that in a slightly more complicated model, the collapse of the universe was very different from the expansion. I realized that I had made a mistake. In fact, the no boundary condition implied that disorder would continue to increase during the contraction. The thermodynamic and psychological arrows of time would not reverse when the universe begins to recontract or inside black holes.

What should you do when you find you have made a mistake like that? Some people, like Eddington, never admit that they are wrong. They continue to find new, and often mutually inconsistent, arguments to support their case. Others claim to have never really supported the incorrect view in the first place or, if they did, it was only to show that it was inconsistent. I could give a large number of examples of this, but I won't because it would make me too unpopular. It seems to me much better and less confusing if you admit in print that you were wrong. A good example of this was Einstein, who said that the cosmological constant, which he introduced when he was trying to make a static model of the universe, was the biggest mistake of his life.

LECTURE 7
THE THEORY OF EVERYTHING

It would be very difficult to construct a complete unified theory of everything all at one go. So instead we have made progress by finding partial theories. These describe a limited range of happenings and neglect other effects, or approximate them by certain num bers. In chemistry, for example, we can calculate the interactions of atoms without knowing the internal structure of the nucleus of an atom. Ultimately, however, one would hope to find a complete, con-sistent, unified theory that would include all these partial theories as approximations. The quest for such a theory is known as "the unification of physics."

Einstein spent most of his later years unsuccessfully searching for a unified theory, but the time was not ripe: Very little was known about the nuclear forces. Moreover, Einstein refused to believe in the reality of quantum mechanics, despite the important role he had played in its development. Yet it seems that the uncertainty principle is a fundamental feature of the universe we live in. A successful unified theory must therefore necessarily incorporate this principle.

The prospects for finding such a theory seem to be much better now because we know so much more about the universe. But we must beware of overconfidence. We have had false dawns before. At the beginning of this century, for example, it was thought that everything could be explained in terms of the properties of continuous matter, such as elasticity and heat conduction. The discovery of atomic structure and the uncertainty principle put an end to that.

Then again, in 1928, Max Born told a group of visitors to Göttingen University, "Physics, as we know it, will be over in six months." His confidence was based on the recent discovery by Dirac of the equation that governed the electron. It was thought that a similar equation would govern the proton, which was the only other particle known at the time, and that would be the end of theoretical physics. However, the discovery of the neutron and of nuclear forces knocked that one on the head, too.

Having said this, I still believe there are grounds for cautious optimism that we may now be near the end of the search for the ultimate laws of nature. At the moment, we have a number of partial theories. We have general relativity, the partial theory of gravity, and the partial theories that govern the weak, the strong, and the electromagnetic forces. The last three may be combined in so-called grand unified theories. These are not very satisfactory because they do not include gravity. The main difficulty in finding a theory that unifies gravity with the other forces is that general relativity is a classical theory. That is, it does not incorporate the uncertainty principle of quantum mechanics. On the other hand, the other partial theories depend on quantum mechanics in an essential way. A necessary first step, therefore, is to combine general relativity with the uncertainty principle. As we have seen, this can produce some remarkable consequences, such as black holes not being black, and the universe being completely self-contained and without boundary. The trouble is, the uncertainty principle means that even empty space is filled with pairs of virtual particles and antiparticles. These pairs would have an infinite amount of energy. This means that their gravitational attraction would curve up the universe to an infinitely small size.

Rather similar, seemingly absurd infinities occur in the other quantum theories. However, in these other theories, the infinities can be canceled out by a process called renormalization. This involves adjusting the masses of the particles and the strengths of the forces in the theory by an infinite amount. Although this technique is rather dubious mathematically, it does seem to work in practice. It has been used to make predictions that agree with observations to an extraordinary degree of accuracy. Renormalization, however, has a serious drawback from the point of view of trying to find a complete theory. When you subtract infinity from infinity, the answer can be anything you want. This means that the actual values of the masses and the strengths of the forces cannot be predicted from the theory. Instead, they have to be chosen to fit the observations. In the case of general relativity, there are only two quantities that can be adjusted: the strength of gravity and the value of the cosmological constant. But adjusting these is not sufficient to remove all the infinities. One therefore has a theory that seems to predict that certain quantities, such as the curvature of space-time, are really infinite, yet these quantities can be observed and measured to be perfectly finite. In an attempt to overcome this problem, a theory called "supergravity" was suggested in 1976. This theory was really just general relativity with some additional particles.

In general relativity, the gravitational force can be thought of as being carried by a particle of spin 2 called the graviton. The idea was to add certain other new particles of spin 3/2, 1, 1/2, and 0. In a sense, all these particles could then be regarded as different aspects of the same "superparticle." The virtual particle/antiparticle pairs of spin 1/2 and 3/2 would have negative energy. This would tend to cancel out the positive energy of the virtual pairs of particles of spin 0, 1, and 2. In this way, many of the possible infinities would cancel out, but it was suspected that some infinities might still remain. However, the calculations required to find out whether there were any infinities left uncanceled were so long and difficult that no one was prepared to undertake them. Even with a computer it was reckoned it would take at least four years. The chances were very high that one would make at least one mistake, and probably more. So one would know one had the right answer only if someone else repeated the calculation and got the same answer, and that did not seem very likely.

Because of this problem, there was a change of opinion in favor of what are called string theories. In these theories the basic objects are not particles that occupy a single point of space. Rather, they are things that have a length but no other dimension, like an infinitely thin loop of string. A particle occupies one point of space at each in stant of time. Thus, its history can be represented by a line in space-time called the "world-line." A string, on the other hand, occupies a line in space at each moment of time. So its history in space-time is a two-dimensional surface called the "world-sheet." Any point on such a world-sheet can be described by two numbers, one specifying the time and the other the position of the point on the string. The world-sheet of a string is a cylinder or tube. A slice through the tube is a circle, which represents the position of the string at one particular time.

Two pieces of string can join together to form a single string. It is like the two legs joining on a pair of trousers. Similarly, a single piece of string can divide into two strings. In string theories, what were previously thought of as particles are now pictured as waves traveling down the string, like waves on a washing line. The emission or absorption of one particle by another corresponds to the dividing or joining together of strings. For example, the gravitational force of the sun on the Earth corresponds to an H-shaped tube or pipe. String theory is rather like plumbing, in a way. Waves on the two vertical sides of the H correspond to the particles in the sun and the Earth, and waves on the horizontal crossbar correspond to the gravitational force that travels between them.

String theory has a curious history. It was originally invented in the late 1960s in an attempt to find a theory to describe the strong force. The idea was that particles like the proton and the neutron could be regarded as waves on a string. The strong forces between the particles would correspond to pieces of string that went between other bits of string, like in a spider's web. For this theory to give the observed value of the strong force between particles, the strings had to be like rubber bands with a pull of about ten tons.

In 1974 Joël Scherk and John Schwarz published a paper in which they showed that string theory could describe the gravitational force, but only if the tension in the string were very much higher—about 10^{39}tons. The predictions of the string theory would be just the same as those of general relativity on normal length scales, but they would differ at very small distances—less than 10^{-33} centimeters. Their work did not receive much attention, however, because at just about that time, most people abandoned the original string theory of the strong force. Scherk died in tragic circumstances. He suffered from diabetes and went into a coma when no one was around to give him an injection of insulin. So Schwarz was left alone as almost the only supporter of string theory, but now with a much higher proposed value of the string tension.

There seemed to have been two reasons for the sudden revival of interest in strings in 1984. One was that people were not really making much progress toward showing that supergravity was finite or that it could explain the kinds of particles that we observe. The other was the publication of a paper by John Schwarz and Mike Green which showed that string theory might be able to explain the existence of particles that have a built-in left-handedness, like some of the particles that we observe. Whatever the reasons, a large number of people soon began to work on string theory. A new version was developed, the so-called heterotic string. This seemed as if it might be able to explain the types of particle that we observe.

String theories also lead to infinities, but it is thought they will all cancel out in versions like the heterotic string. String theories, however, have a bigger problem. They seem to be consistent only if space-time has either ten or twenty-six dimensions, instead of the usual four. Of course, extra space-time dimensions are a commonplace of science fiction; indeed, they are almost a necessity. Otherwise, the fact that relativity implies that one cannot travel faster than light means that it would take far too long to get across our own galaxy, let alone to travel to other galaxies. The science fiction idea is that one can take a shortcut through a higher dimension. One can picture this in the following way. Imagine that the space we live in had only two dimensions and was curved like the surface of a doughnut or a torus. If you were on one side of the ring and you wanted to get to a point on the other side, you would have to go around the ring. However, if you were able to travel in the third dimension, you could cut straight across.

Why don't we notice all these extra dimensions if they are really there? Why do we see only three space and one time dimension? The suggestion is that the other dimensions are curved up into a space of very small size, something like a million million million million millionth of an inch. This is so small that we just don't notice it. We see only the three space and one time dimension in which space-time is thoroughly flat. It is like the surface of an orange: if you look at it close up, it is all curved and wrinkled, but if you look at it from a distance, you don't see the bumps and it appears to be smooth. So it is with space-time. On a very small scale, it is ten-dimensional and highly curved. But on bigger scales, you don't see the curvature or the extra dimensions.

If this picture is correct, it spells bad news for would-be space travelers. The extra dimensions would be far too small to allow a spaceship through. However, it raises another major problem. Why should some, but not all, of the dimensions be curled up into a small ball? Presumably, in the very early universe, all the dimensions would have been very curved. Why did three space and one time dimension flatten out, while the other dimensions remained tightly curled up?

One possible answer is the anthropic principle. Two space dimensions do not seem to be enough to allow for the development of complicated beings like us. For example, two-dimensional people living on a one-dimensional Earth would have to climb over each other in order to get past each other. If a twodimensional creature ate something it could not digest completely, it would have to bring up the remains the same way it swallowed them, because if there were a passage through its body, it would divide the creature into two separate parts. Our two-dimensional being would fall apart. Similarly, it is difficult to see how there could be any circulation of the blood in a two-dimensional creature. There would also be problems with more than three space dimensions. The gravitational force between two bodies would decrease more rapidly with distance than it does in three dimensions. The significance of this is that the orbits of planets, like the Earth, around the sun would be unstable. The least disturbance from a circular orbit, such as would be caused by the gravitational attraction of other planets, would cause the Earth to spiral away from or into the sun. We would either freeze or be burned up. In fact, the same behavior of gravity with distance would mean that the sun would also be unstable. It would either fall apart or it would collapse to form a black hole. In either case, it would not be much use as a source of heat and light for life on Earth. On a smaller scale, the electrical forces that cause the electrons to orbit around the nucleus in an atom would behave in the same way as the gravitational forces. Thus, the electrons would either escape from the atom altogether or it would spiral into the nucleus. In either case, one could not have atoms as we know them.

It seems clear that life, at least as we know it, can exist only in regions of space-time in which three space and one time dimension are not curled up small. This would mean that one could appeal to the anthropic principle, provided one could show that string theory does at least allow there to be such regions of the universe. And it seems that indeed each string theory does allow such regions. There may well be other regions of the universe, or other universes (whatever that may mean) in which all the dimensions are curled up small, or in which more than four dimensions are nearly flat. But there would be no intelligent beings in such regions to observe the different number of effective dimensions.

Apart from the question of the number of dimensions that space-time appears to have, string theory still has several other problems that must be solved before it can be acclaimed as the ultimate unified theory of physics. We do not yet know whether all the infinities cancel each other out, or exactly how to relate the waves on the string to the particular types of particle that we observe. Nevertheless, it is likely that answers to these questions will be found over the next few years, and that by the end of the century we shall know whether string theory is indeed the long sought-after unified theory of physics.

Can there really be a unified theory of everything? Or are we just chasing a mirage? There seem to be three possibilities:

- There really is a complete unified theory, which we will someday discover if we are smart enough.
- There is no ultimate theory of the universe, just an infinite sequence of theories that describe the universe more and more accurately.
- There is no theory of the universe. Events cannot be predicted beyond a certain extent but occur in a random and arbitrary manner.

Some would argue for the third possibility on the grounds that if there were a complete set of laws, that would infringe on God's freedom to change His mind and to intervene in the world. It's a bit like the old paradox: Can God make a stone so heavy that He can't lift it? But the idea that God might want to change His mind is an example of the fallacy, pointed out by St. Augustine, of imagining God as a being existing in time. Time is a property only of the universe that God created. Presumably, He knew what He intended when He set it up.

With the advent of quantum mechanics, we have come to realize that events cannot be predicted with complete accuracy but that there is always a degree of uncertainty. If one liked, one could ascribe this randomness to the intervention of God. But it would be a very strange kind of intervention. There is no evidence that it is directed toward any purpose. Indeed, if it were, it wouldn't be random. In modern times, we have effectively removed the third possibility by redefining the goal of science. Our aim is to formulate a set of laws that will enable us to predict events up to the limit set by the uncertainty principle.

The second possibility, that there is an infinite sequence of more and more refined theories, is in agreement with all our experience so far. On many occasions, we have increased the sensitivity of our measurements or made a new class of observations only to discover new phenomena that were not predicted by the existing theory. To account for these, we have had to develop a more advanced theory. It would therefore not be very surprising if we find that our present grand unified theories break down when we test them on bigger and more powerful particle accelerators. Indeed, if we didn't expect them to break down, there wouldn't be much point in spending all that money on building more powerful machines.

However, it seems that gravity may provide a limit to this sequence of "boxes within boxes." If one had a particle with an energy above what is called the Planck energy, 1019 GeV, its mass would be so concentrated that it would cut itself off from the rest of the universe and form a little black hole. Thus, it does seem that the sequence of more and more refined theories should have some limit as we go to higher and higher energies. There should be some ultimate theory of the universe. Of course, the Planck energy is a very long way from the energies of around a GeV, which are the most that we can produce in the laboratory at the present time. To bridge that gap would require a particle accelerator that was bigger than the solar system. Such an accelerator would be unlikely to be funded in the present economic climate.

However, the very early stages of the universe are an arena where such energies must have occurred. I think that there is a good chance that the study of the early universe and the requirements of mathematical consistency will lead us to a complete unified theory by the end of the century-always presuming we don't blow ourselves up first.

What would it mean if we actually did discover the ultimate theory of the universe? It would bring to an end a long and glorious chapter in the history of our struggle to understand the universe. But it would also revolutionize the ordinary person's understanding of the laws that govern the universe. In Newton's time it was possible for an educated person to have a grasp of the whole of human knowledge, at least in outline. But ever since then, the pace of development of science has made this impossible. Theories were always being changed to account for new observations. They were never properly digested or simplified so that ordinary people could understand them.You had to be a specialist, and even then you could only hope to have a proper grasp of a small proportion of the scientific theories.

Further, the rate of progress was so rapid that what one learned at school or university was always a bit out of date. Only a few people could keep up with the rapidly advancing frontier of knowledge. And they had to devote their whole time to it and specialize in a small area. The rest of the population had little idea of the advances that were being made or the excitement they were generating.

Seventy years ago, if Eddington is to be believed, only two people understood the general theory of relativity. Nowadays tens of thousands of university graduates understand it, and many millions of people are at least familiar with the idea. If a complete unified theory were discovered, it would be only a matter of time before it was digested and simplified in the same way. It could then be taught in schools, at least in outline. We would then all be able to have some understanding of the laws that govern the universe and which are responsible for our existence.

Einstein once asked a question: "How much choice did God have in constructing the universe?" If the no boundary proposal is correct, He had no freedom at all to choose initial conditions. He would, of course, still have had the freedom to choose the laws that the universe obeyed. This, however, may not really have been all that much of a choice. There may well be only one or a small number of complete unified theories that are self-consistent and which allow the existence of intelligent beings.

We can ask about the nature of God even if there is only one possible unified theory that is just a set of rules and equations. What is it that breathes fire into the equations and makes a universe for them to describe? The usual approach of science of constructing a mathematical model cannot answer the question of why there should be a universe for the model to describe. Why does the universe go to all the bother of existing? Is the unified theory so compelling that it brings about its own existence? Or does it need a creator, and, if so, does He have any effect on the universe other than being responsible for its existence? And who created Him?

Up until now, most scientists have been too occupied with the development of new theories that describe what the universe is, to ask the question why. On the other hand, the people whose business it is to ask why—the philosophers—have not been able to keep up with the advance of scientific theories. In the eighteenth century, philosophers considered the whole of human knowledge, including science, to be their field. They discussed questions such as: Did the universe have a beginning? However, in the nineteenth and twentieth centuries, science became too technical and mathematical for the philosophers or anyone else, except a few specialists. Philosophers reduced the scope of their inquiries so much that Wittgenstein, the most famous philosopher of this century, said, "The sole remaining task for philosophy is the analysis of language." What a comedown from the great tradition of philosophy from Aristotle to Kant.

However, if we do discover a complete theory, it should in time be understandable in broad principle by everyone, not just a few scientists. Then we shall all be able to take part in the discussion of why the universe exists. If we find the answer to that, it would be the ultimate triumph of human reason. For then we would know the mind of God.